African Insectivora and Elephant-Shrews

An Action Plan for their Conservation

Compiled by
Martin E. Nicoll and Galen B. Rathbun
IUCN/SSC Insectivore, Tree-Shrew and Elephant-Shrew
Specialist Group

Contents

	Page
Foreword..	iii
Acknowledgements..	iii

Chapter 1: Introduction.. 1
1.1 The Insectivora of Africa and Madagascar............... 1
1.2 The Elephant-shrews.. 2
1.3 The IUCN/SSC Insectivore, Tree-shrew and Elephant-shrew Specialist Group.................................. 2
1.4 Threats Facing the African Insectivora and Elephant-shrews.. 2
1.5 Rationale and Objectives of the Action Plan............. 4

Chapter 2: The Insectivora of Madagascar............... 5
2.1 Introduction... 5
2.2 Classification... 5
2.3 Biogeographic Regions of Madagascar..................... 6
2.4 Species of Conservation Concern............................. 9

Chapter 3: The Insectivora of Africa........................ 13
3.1 Introduction... 13
3.2 Classification... 13
3.3 Biogeographic Regions of Africa............................ 14
3.4 Species of Conservation Concern........................... 15

Chapter 4: The Elephant-shrews.............................. 22
4.1 Introduction... 22

	Page
4.2 Classification..	22
4.3 Biogeographic Review of the Elephant-shrews.......	22
4.4 Taxa of Conservation Concern................................	25
4.5 Captive Breeding of Elephant-shrews.....................	27

Chapter 5: Conservation Action Plan....................... 30
5.1 Rationale for Insectivora and Elephant-shrew Conservation... 30
5.2 Habitat Protection.. 30
5.3 Field Surveys... 35
5.4 Research... 41
5.5 Captive Breeding... 42
5.6 Priorities for Action 43

Appendix 1: Conservation Proposal for the Aquatic Tenrec.. 44
Appendix 2: Conservation of the Golden Moles, Especially the Genus *Chrysospalax* 47
Appendix 3: An Evaluation of the Status of the Elephant-shrews in Coastal East Africa.. 48
Appendix 4: Members of the IUCN/SSC Insectivore, Tree-shrew, and Elephant-shrew Specialist Group.. 50

References... 52

Foreword

Extinction of an elephant species will resound around the world; extinction of an insectivore or elephant-shrew will go undetected for decades, and even then few people will care. But somewhere back in the corner of our brains we are totally convinced that this action plan is important. It is important because we believe in the tenets of reverence for life, that biodiversity is central to our survival, and that nobody knows the point at which species extinctions will become so serious that ecosystems are irrepairably damaged. In some ways this document is testimony to biologists' greatest dedication to the animals they study, a last passionate, yet objective, plea: if natural habitats are to be destroyed, at the very least let us consider these precious few areas that are inhabited by the beautiful animals we have learned to treasure for their own sake.

In nature and in zoos most people rarely appreciate the variety of insectivores beyond moles and hedgehogs. Even the hunter and 'woods-wise' person living in a forest village in Africa might never see some of the insectivores that dwell in his area. Yet, in Madagascar, tenrecs form the majority of the small mammals; some of these diminutive species are likely to be safe, except in the event of major disruption to the integrity of the ecosystems they inhabit. On mainland Africa, the diversity of insectivores is often outnumbered by rodents. In all probability, some insectivore species have become extinct before discovery. Despite the disappearance of vast tracts of natural habitat, there remain a huge number of areas in need of surveying, more than any single biologist could accomplish in a lifetime. This action plan provides direction and emphasis in the important task ahead, a job that will be critical in our future attempts to save biodiversity in Africa.

Edwin Gould
Curator of Mammals
National Zoo, Smithsonian Institution

Acknowledgments

This Action Plan could not have been produced without the assistance of a large number of people. We are grateful for the information and guidance we received from the following people: Michel Anciaux de Faveaux, Stephane Aulagnier, Eytan Avital, J. Bernard, J.T. Christensen, Sara Churchfield, M. Colyn, Ken Creighton, John Fanshawe, David Happold, H. Hayaki, Graham Hickman, Graham Kepley, Olivier Langrand, Jeff McNeely, J. Meester, D. Meirte, Jan Nel, F. Petter, John Pettigrew, Felix Rakotondraparany, I.L. Rautenbach, Duane Schlitter, B. Sige, Michel Seguignes, P.J. Stephenson, Chris Stuart, Michel Thevenot, Jacutes Verschuren, Brigitte Wendhold, Lucienne Wilme, Peter Woodall and Chris Raxworthy. Information and comments from W.F.H. Ansell, Jon Lovett, S.A. Robertson, Dave Stone and Tom Struhsaker were especially useful. Fred Koontz contributed information on captive breeding of elephant-shrews. Special mention must be made of the assistance provided by Rainer Hutterer, who carefully reviewed and revised the sections on African shrews. The editorial comments on early drafts of the elephant-shrew section by R.L. Brownell, Jr. and Paul Opler were appreciated. Simon Stuart prepared the document for publication, and provided much valuable advice. The final editing and drawing together of the document was carried out by Ann Stuart, and we are most grateful to her for her hard work.

Martin E. Nicoll
Chairman, IUCN/SSC Insectivore, Tree-Shrew and Elephant-Shrew Specialist Group
WWF Madagascar Programme

Galen B. Rathbun
U.S. Fish & Wildlife Service

The rough-haired golden mole (*Chrysospalax villosus*)
(Photo by G. C. Hickman)

Chapter 1: Introduction

The greater hedgehog-tenrec *Setifer setosus* (Photo by M. E. Nicoll/BIOS)

The three mammalian orders Insectivora (which includes the shrews, moles and hedgehogs), Scandentia (the tree-shrews), and Macroscelidea (the elephant-shrews) all comprise small, terrestrial mammals that are primarily adapted to an insectivorous diet. The Insectivora is by far the largest of these, being represented by some 400 species distributed throughout the world with the exception of most of South America, Australasia, Antarctica, and most oceanic islands. The tree-shrews constitute a small group of 16 species limited to south and south-east Asia. The elephant-shrews are a group of 15 species, all restricted to the African continent. This Action Plan, prepared by the IUCN/SSC Insectivore, Tree-Shrew and Elephant-Shrew Specialist Group (ITSES), addresses the conservation needs of the Insectivora and Macroscelidea of Africa and Madagascar.

1.1 The Insectivora of Africa and Madagascar

The Insectivora includes some of the most primitive groups of placental mammals. Many of them have features and lifestyles that were much more widespread during the early stages of mammalian evolution than they are today. Thus, research on the Insectivora has provided considerable insight into how mammals have evolved from reptilian ancestors, and has also shed light on how some of our early mammalian ancestors lived. The shrews, hedgehogs and tenrecs, all represented in Africa or Madagascar, have been most instructive in this respect. For example, research on the tenrecs has helped clarify how some of the driving evolutionary changes, such as raised body temperature, have been fundamental processes in the evolution of mammals.

The Insectivora comprise a wide variety of largely insectivorous mammals which are often grouped together because they have common morphological traits rather than a clear, recent common ancestry. This is true for Africa and Madagascar where four disparate families occur: the true hedgehogs (family Erinaceidae) are far removed from the shrews (Soricidae), and both are only distantly related to either the tenrecs (Tenrecidae) and the golden moles (Chrysochloridae). The tenrecs and golden moles, however, are sufficiently similar to enable them to be grouped together into the sub-order Tenre-

The rufous elephant-shrew (*Elephantus rufescens*)
(Photo by G. B. Rathbun)

comorpha, reflecting a common, though remote, ancestry. The number of species in each of these four families, that are known from various African countries, is shown in Table 1.1.

The shrew and hedgehog families are well represented on continents other than Africa, but as many as one-third of all shrews occur only in Africa. The golden moles and tenrecs are entirely confined to the African and Malagasy regions respectively. Both the morphological characteristics and distribution patterns suggest that golden moles and tenrecs are the surviving representatives of ancient southern continental mammalian radiations. The golden moles are largely confined to southern Africa and no species penetrates the Saharan zone to the north. Of the tenrecs, the entire family is restricted to Madagascar, except for three specialized aquatic species occurring in rainforest or montane regions in continental Africa.

1.2 The Elephant-shrews

Elephant-shrews form a cohesive group of 15 species confined entirely to Africa. They take their name from their long mobile snouts. For a long time, they were classified as the Family Macroscelididae within the Order Insectivora, even though they share few common traits with other Insectivora except for eating insects. Now, this group is recognized as a full Order, the Macroscelidea, having a long history of separation from other mammalian groups. Indeed, the closest ancestors may be rabbits and hares, but their common ancestry is obviously remote. Table 1.1 shows the number of elephant-shrew species known from various African countries.

Elephant-shrews are largely crepuscular or diurnal, and insectivorous. For small mammals, with a weight range of 45 - 550 g, they have several unusual and evolutionarily interesting life history traits. Adults have large rear legs (thus their name, *macro* + *scelidea* = big thigh) that they use for swift, cursorial locomotion. The young are born in a well-developed state, which enables them to run nearly as fast as adults within hours or days of birth. Another unusual trait is that some species form territories that a monogamous pair defends against other elephant-shrews. All of these characters are very unlike those found in most other small mammals, especially the Insectivora.

1.3 The IUCN/SSC Insectivore, Tree-Shrew and Elephant-Shrew Specialist Group

The IUCN Species Survival Commission now has 90 Specialist Groups covering a wide variety of taxonomic groups and conservation techniques. The common objective of all these groups is to provide technical advice and guidance to the world conservation community (usually taking the form of the member organisations and governments of IUCN) on the conservation of the species within their brief. One of the main activities of all the taxa-focussed Specialist Groups is to prepare Action Plans which outline the conservation priorities for their species. The present document covering Africa is the first in a series of regional Action Plans being prepared by the Insectivore, Tree-Shrew and Elephant-Shrew Specialist Group (ITSES). It will be followed by similar documents covering tropical Asia, Europe and northern Asia, and the Americas.

Although the Species Survival Commission has carried out work on the Insectivora and elephant-shrews for some time, the Specialist Group in its current form was re-established in 1986. A list of the current members is included as Appendix 2. ITSES has already been successful in promoting conservation action on behalf of threatened species. The most notable project so far promoted by the Group is that on the conservation of the aquatic tenrec (*Limnogale mergulus*). This project proposal is included as Appendix 1 of the Action Plan. As a result of this Action Plan, the Group intends to promote many more projects along these lines. The Group also produces an occasional newsletter which aims to share information among people working on the conservation of these poorly known species.

1.4 Threats facing the African Insectivora and Elephant-shrews

In general, we know very little about the threats facing the African Insectivora and elephant-shrews. However, the little we do know suggests that they tend to be at risk from habitat modification, rather than from direct exploitation of the species themselves. Both on the African mainland and on Madagascar there are several centres of exceptional diversity and endemism for these species. These centres are identified in this Action Plan, and the principal conservation priority is the maintenance of sufficient areas of suitable habitat in these centres. These same areas are also known to be centres of diversity for other species, such as primates, rodents, birds, amphibians and plants, and so their survival is an important global priority for the conservation of biological diversity.

A juvenile tail-less tenrec (*Tenrec ecaudatus*)
(Photo by M. E. Nicoll/BIOS)

Table 1.1: The number of species of Insectivora and elephant-shrews known from selected African countries. Note that the various families of Insectivora are shown separately.

Country	Surface Area (km²)	Shrews	Hedgehogs	Golden Moles	Tenrecs and Otter-shrews	Elephant-shrews	Total
Algeria	2,381,741	5	2	0	0	1	8
Angola	1,246,700	13	1	1	1	3	19
Botswana	600,372	8	1	0	0	4	13
Burkina Faso	274,200	11	1	0	0	0	12
Cameroon	475,442	38	1	1	1	0	41
Cent. Afr. Republic	624,977	?	1	0	1	1	3+
Chad	1,284,000	5	2	0	0	0	7
Congo	342,000	?	0	0	1	2	3+
Egypt	997,667	5	2	0	0	0	7
Ethiopia	1,221,900	20	2	0	0	1	23
Gabon	267,667	14	0	0	1	0	15
Gambia	11,295	6	0	0	0	0	6
Ghana	238,537	16	1	0	0	0	17
Guinea	245,857	16	1	0	1	0	18
Ivory Coast	322,468	18	1	0	1	0	20
Kenya	582,600	31	1	0	1	5	38
Liberia	111,369	10	0	0	1	0	11
Libya	1,759,640	2	3	0	0	1	6
Madagascar	594,180	2	0	0	24	0	26
Malawi	118,484	10	2	0	0	4	16
Mali	1,204,021	9	2	0	0	0	11
Morocco	458,730	6	2	0	0	1	9
Mauritania	1,030,700	5	1	0	0	0	6
Mozambique	784,961	10	2	1	0	6	19
Namibia	823,169	6	1	1	0	5	13
Niger	1,267,000	8	1	0	0	0	9
Nigeria	923,768	24	1	0	1	0	26
Rwanda	26,338	15	1	1	0	1	18
Sao Tome & Principe	964	1	0	0	0	0	1
Senegal	196,772	8	1	0	0	0	9
Sierra Leone	71,740	16	1	0	0	0	17
Somalia	637,657	9	2	1	0	2	14
South Africa	1,221,037	18	1	14	0	8	41
Sudan	2,505,813	21	2	0	0	3	26
Tanzania	945,087	27	1	1	0	5	34
Togo	56,000	13	1	0	0	0	14
Tunisia	164,150	3	1	0	0	1	5
Uganda	241,139	30	1	1	0	4	36
Western Sahara	266,000	3	1	0	0	0	4
Zaire	2.345,309	56	1	2	2	4	65
Zambia	752,614	13	2	0	1	4	20
Zimbabwe	390,272	11	1	1	0	4	17

The increasing loss and fragmentation of remaining natural environments (in particular forests) over much of the African mainland and Madagascar constitute a serious threat to some of the world's richest centres of biological diversity, as well as spelling potential doom to a considerable number of Insectivora and elephant-shrew taxa. In most cases, the driving force to clear habitats is the need for more agricultural land for local people. The long-term future therefore depends on the establishment of protected areas within the context of integrated rural development programmes that aim to provide direct benefits to local people as a result of conservation efforts. Habitats are also threatened by larger-scale commercial schemes, the most notorious being the devastation of Mount Nimba in Liberia by bauxite mining operations. This particular example of unsustainable development is a model of the sort of exploitation that should be avoided.

In general, all species that cannot adapt to man-made environments are potentially at risk. This risk is obviously greater where the geographical range of a species is smaller, especially for a species restricted to habitat that is being converted or is unprotected. The aquatic species (i.e. the three species of otter-shrew and the aquatic tenrec) are also susceptible to pollution and the siltation of rivers resulting from soil erosion, and so are particularly vulnerable. The consequences of local extirpation of aquatic species can be particularly serious, since they generally find it more difficult to recolonise areas, especially if they are eradicated from entire river systems.

Direct exploitation of species is limited. Giant otter-shrews are hunted for the pelts, and some hedgehog, tenrec and elephant-shrew species are eaten, but there is little evidence to suggest that this is threatening.

1.5 Rationale and Objectives of the Action Plan

Threats to large spectacular animals or attractive plants are often quickly recognized by biologists, and support can often be quickly generated among organisations or governments that are in a position to remedy the problem. For the far greater numbers of less conspicuous plants or animals, however, the threats often go unnoticed, and raising interest and support for remedial action can be more difficult. This is the case for most of the Insectivora and elephant-shrews.

Except for the European and North American species, very little is known about the Insectivora. This is partly a result of their cryptic habits and because comparatively few attempts have been made to study them, but it may also be due to their rarity or their small ranges. Many of the African and Malagasy Insectivora are known from only one or a few specimens, indicating that little basic survey work has been carried out. Recent field studies have even revealed new species. Many of the little known or newly discovered species are found within the centres of greatest Insectivora species richness, in particular the central African and eastern Malagasy rainforests, often occurring in seriously threatened habitats. Some Insectivora in Rwanda and Madagascar, for example, occur only in natural moist forests and cannot tolerate serious habitat disturbance. These species may be at considerable risk as the native forests are cleared for cultivation and the remaining suitable habitats become smaller and increasingly fragmented.

The elephant-shrews are better known than most African or Malagasy Insectivora, but only three species have been reasonably well studied. Some are known to be at risk and to have need of protection either through conservation in the wild or in zoological parks. For these threatened species and those that are less well known, much more information is needed in order to ensure that this ancient and curious group suffers no extinctions.

This Action Plan covers the entire African continent and its in-shore islands, together with the island-continent of Madagascar, and deals with all members of the mammalian Orders Insectivora and Macroscelidea that occur within this geographical area.

It is important to have an Action Plan for Insectivora and elephant-shrews for several reasons. The objectives of the Action Plan are to:

- Emphasise the often overlooked importance of the Insectivora and Macroscelidea, either as important elements in many ecosystems, or as ancient lineages which are useful in elucidating the mammalian evolutionary process;

- Update the knowledge on the status of uncommon species;

- Identify species and key habitats at risk;

- Identify specific field conservation projects that are needed for these species;

- Identify which of the threatened species require captive breeding;

- Encourage wider participation in conservation projects for these little known species;

- Strengthen arguments for the protection of key threatened ecosystems, especially in areas of high biological diversity and endemism.

Chapter 2: The Insectivora of Madagascar

2.1 Introduction

Madagascar has probably been isolated from mainland Africa for at least 160 million years. Although less than 400 km from the East African coast, the impact of this isolation has been marked, with a highly distinctive floral and faunal assemblage containing many lineages that have long since disappeared on the African mainland. Owing to the island's long history of isolation and its highly varied geomorphology, vegetation and climate, many Malagasy animal families have undergone spectacular adaptive radiations to occupy niches which, on the contiguous continental land masses, would be occupied by several divergent families. The occurrence of relatively primitive lineages with marked diversity means that Madagascar is one of the most interesting living laboratories of evolution. Given its biological value, and because all of its terrestrial ecosystems are currently under severe threat, Madagascar is one of the world's highest conservation priorities.

The Madagascar pygmy shrew (*Suncus madagascariensis*)
(Photo by M. E. Nicoll/BIOS)

2.2 Classification

Two groups of the Insectivora are present in Madagascar, the Tenrecidae (tenrecs) and Soricidae (shrews). Just two species of shrew are present: the musk shrew (*Suncus murinus*), and the Madagascar pygmy shrew (*S. madagascariensis*). The musk shrew is a species commensal with man and was certainly introduced from Asia, while the Madagascar pygmy shrew is also likely to be a recent colonist or introduction. It is probably the same species as the pygmy white-toothed shrew (*S. etruscus*) from Africa and Europe. Both species of shrew on Madagascar are widespread and common, and neither requires conservation measures.

The Malagasy tenrecs are the country's oldest surviving mammalian lineage and the most diversified group of living Insectivora on the island. All the species on Madagascar are endemic, and there are only three other species elsewhere, the otter-shrews (which occur in Africa where the family was apparently once more widespread and diversified). The Malagasy Tenrecidae comprise two divergent groups, the spiny tenrecs in the sub-family Tenrecinae, and the furred tenrecs in the sub-family Oryzorictinae (see Table 2.1).

The furred tenrecs (Oryzorictinae) are highly diversified, occupying niches similar to those elsewhere of shrews, shrew-moles, aquatic otter-shrews (the African tenrec group), and aquatic desmans. Many species are known from only one or a few specimens, and the taxonomy of the largest genus, *Microgale* (the shrew-tenrecs), is confused. Because of this confusion, the estimated number of species has ranged from 14 (MacPhee 1987) to 24 (Meester and Setzer 1974). The review by MacPhee (1987) recognized that the previous classifications of *Microgale* species included several forms which should be classed as a single species. However, MacPhee (1987) admitted that his new classification might be over-simplified, since it describes new groupings that may represent "ecological adaptation" syndromes rather than real species differences. New information confirms that some of his new "species" do indeed include several distinct taxa. For these reasons, this Action Plan will use the older classification until further information is available (see Table 2.1). An additional species described recently (*Microgale pulla*: Jenkins 1988) is also included in this table. A new revised taxonomy of the tenrecs, based on recent fieldwork, is included in Table 2.2, though it is recognised that much more research is needed before these problems can be fully resolved.

The spiny tenrecs (Tenrecinae) contain five species in four genera. They have a weight range of 80-2,000g. Members of this group occur in all terrestrial environments. The greater hedgehog-tenrec (*Setifer setosus*) and the lesser hedgehog-tenrec (*Echinops telfairi*) occupy niches which are filled elsewhere by hedgehogs. The tail-less tenrec (*Tenrec ecaudatus*) occupies a niche which is normally filled by moonrats or opossums. The two streaked tenrecs *Hemicentetes nigriceps* and *H. semispinosus* have life styles that are not found among

A shrew-tenrec (*Oryzorictes talpoides*)
(Photo by M. E. Nicoll/BIOS)

Table 2.1: Taxonomy of the tenrecs (family Tenrecidae), showing a comparison between the different classifications of shrew-tenrecs (*Microgale* spp) adopted by Meester and Setzer (1974) and MacPhee (1987).

Meester and Setzer (1974)	MacPhee (1987)
Furred Tenrecs: Subfamily Oryzorictinae (Madagascar)	
Limnogale mergulus	-
Oryzorictes talpoides	-
O. hova	-
O. tetradactylus	-
Geogale aurita	-
Microgale longicaudata	*M. longicaudata*
M. majori	syn *M. longicaudata*
M. prolixacaudata	syn *M. longicaudata*
M. cowani	*M. cowani*
M. crassipes	syn *M. cowani*
M. longirostris	syn *M. cowani*
M. taiva	syn *M. cowani*
M. drouhardi	syn *M. cowani*
M. melanorrachis	syn *M. cowani*
M. dobsoni	*M. dobsoni*
M. thomasi	*M. thomasi*
M. talazaci	*M. talazaci*
M. gracilis	*M. gracilis*
M. pusilla	*M. pusilla*
M. brevicaudata	*M. brevicaudata*
M. breviceps (extinct)	syn *M. brevicaudata*
M. (Paramicrogale) occidentalis	syn *M. brevicaudata*
M. principula	*M. principula*
M. sorella	syn *M. principula*
M. decaryi	syn *M. principula*
M. parvula	*M. parvula*
M. pulla (Jenkins, 1988)	-
Spiny Tenrecs: Subfamily Tenrecinae (Madagascar)	
Tenrec ecaudatus	-
Hemicentetes semispinosus	-
H. nigriceps	-
Setifer setosus	-
Echinops telfairi	-
Otter-shrews: Subfamily Potamogalinae (Africa)	
Potamogale velox	-
Micropotamogale ruwenzorii	-
M. lamottei	-

Table 2.2: Proposed revised taxonomy of the tenrecs (family Tenrecidae) (based on Table 2.1 and field data from M. E. Nicoll, F. Rakotondraparany and P. J. Stephenson).

Furred Tenrecs:
Subfamily Oryzorictinae (Madagascar)

Limnogale mergulus
Oryzorictes talpoides
O. hova
O. tetradactylus
Geogale aurita
Microgale longicaudata
M. cowani
M. longirostris
M. taiva
M. melanorrachis
M. dobsoni
M. thomasi
M. talazaci
M. gracilis
M. pusilla
M. brevicaudata
M. principula
M. parvula
M. pulla

Spiny Tenrecs:
Subfamily Tenrecinae (Madagascar)

Tenrec ecaudatus
Hemicentetes semispinosus
H. nigriceps
Setifer setosus
Echinops telfairi

Otter-shrews:
Subfamily Potamogalinae (Africa)

Potamogale velox
Micropotamogale ruwenzorii
Micropotamogale lamottei

2.3 Biogeographic Regions of Madagascar

In Madagascar, the tenrecs occur in a range of distinctive ecosystems. The different ecosystems are a product of the effects of latitude, prevailing winds, geomorphology and topography. In general there is a trend for the climate to become drier as one moves from east to west owing to the high central plateau blocking moist easterly winds. There is a strong tendency for lower rainfall in the south, with the extreme southwest being the driest region, approaching desert conditions. Two floristic Regional Centres of Endemism are recognized: the East and the West (White 1983). The Eastern Regional Centre of Endemism has four recognized floristic domains:

The lesser hedgehog-tenrec (*Echinops telfairi*)
(Photo by M. E. Nicoll/BIOS)

other members of the Insectivora, and so exhibit unique morphological and anatomical traits.

Since all Malagasy terrestrial or freshwater ecosystems are threatened, any tenrec species that cannot adapt to man-made environments is at risk. All of the spiny tenrecs can adapt to man-made environments and hence are not at risk. However, for the majority of furred tenrecs, the degree of dependence on natural environments is not yet known. Some furred tenrecs also appear to have restricted habitat requirements or ranges, and so it is these species that are considered to be facing real or potential risks.

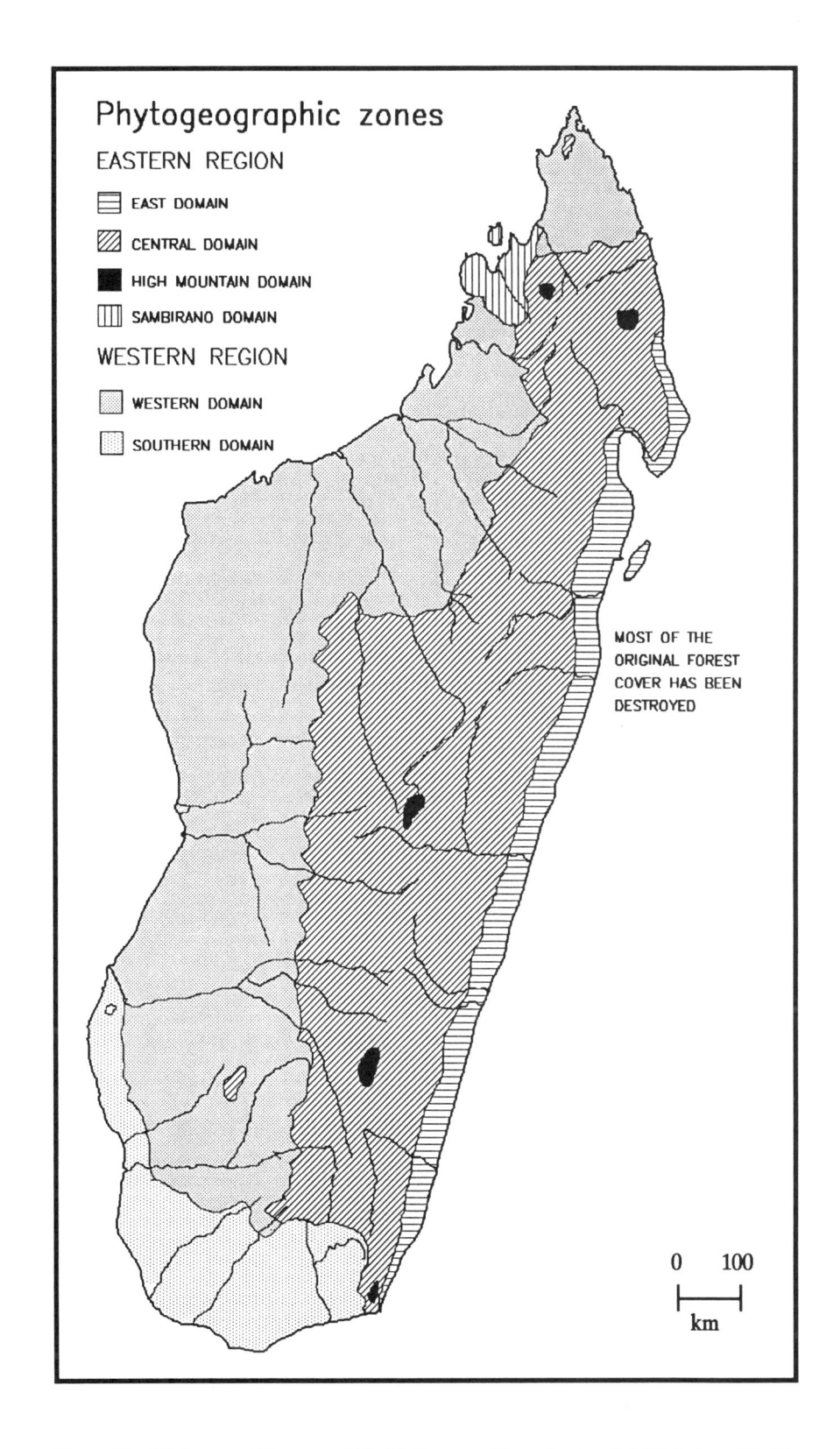

Figure 2.1 Phytogeographic zones within Madagascar. Current research is likely to modify this classification, but the basic separation into Eastern and Western Regions will be retained. Estimations of the extent of foret cover before the arrival of people vary from uninterrupted cover to a mosaic of forest and woodland/savannah, particularly on the central plateau. Approximately 80% of the original cover has disappeared, with virtually none remaining in the centre.

Eastern, Central, High Mountain, and Sambirano; and the Western Centre of Endemism has two domains: the Western and Southern Domains (see Fig. 2.1).

In the Eastern Centre of Endemism, eastern lowland rainforest is the natural vegetation between sea level and 800 m altitude from the extreme south-east northwards to approximately 14°S in the north-east. Most has been cleared, with the few remaining large blocks being in the north-east. Above 800m in the Eastern Regional Centre of Endemism, the Central Domain extends as far west as the 800 m contour on the western edge of the central plateau. Most rainforest has disappeared but this remains the most extensive formation. A band runs discontinuously through the centre-east of the island, extending across to the west in the northern part of the island. Five High Mountain Domains are recognized: Tsaratanana in the north-west, Marojejy in the north-east, Ankaratra in the centre east, Andringitra in the centre-south-east, and Andohahela in the south-east. The natural vegetation of Ankaratra has almost entirely disappeared, and Marojejy and Andohahela are the only intact blocks. The Sambirano Domain forms a small enclave in the north-west and is severely threatened.

In the Western Centre of Endemism, the intact vegetation in the Western Domain remains as discontinuous patches of forest on the coastal plains and limestone plateaux, over much of the west and north. It is severely threatened by clearance for agriculture and by fire. The Southern Domain, is, perhaps, the least disturbed area, but is increasingly being threatened by fire, cultivation, overgrazing, and charcoal production. All domains show localised variation in physiognomy and species composition.

Tenrec diversity is highest in the Eastern Centre of Endemism, but in neither Centre does distribution of tenrecs correlate closely with the different floristic domains. All but one or two of the furred tenrecs occur in the east, while only three or four are known from the west. The east supports 15 species of conservation concern, but these may include several forms which are actually the same species. All but a few of the eastern furred tenrecs appear to require undisturbed or relatively unperturbed environments, in most cases forests, and their ranges are often restricted and isolated. Many of these species may therefore be considered to be under threat from habitat destruction. Further research may reveal that others are threatened even though they enjoy relatively large distributional ranges. Research might also identify new species occurring in restricted areas; one new species has been described during the last three years (Jenkins 1988). Of the spiny tenrecs, all occur in the east except for the lesser hedgehog-tenrec. Two of the spiny tenrec species are confined to the east, but none appears to be threatened.

Three or four furred tenrecs have been recorded in the Western Centre of Endemism. One is widespread in the east and, although known from only one south-western locality and near the centre west, is not apparently threatened and may occur more widely in the west and south. The large-eared tenrec subspecies (*Geogale aurita aurita*) is widespread in the southern sector of the west and in the south, but may be only locally common and is likely to be threatened by forest clearance. The western short-tailed shrew-tenrec (*Microgale (Paramicrogale) occidentalis*) is only known from a small area in the north-west, but its precise localities and habitat are unknown. MacPhee (1987) considers this form to be the same species as the short-tailed shrew-tenrec (*M. brevicaudata*), and specimens fitting the descriptions of these two species have recently been recorded in the centre-west near Morondava.

The three species of the spiny tenrec which occur within the Western Centre of Endemism are widespread and do not require conservation attention.

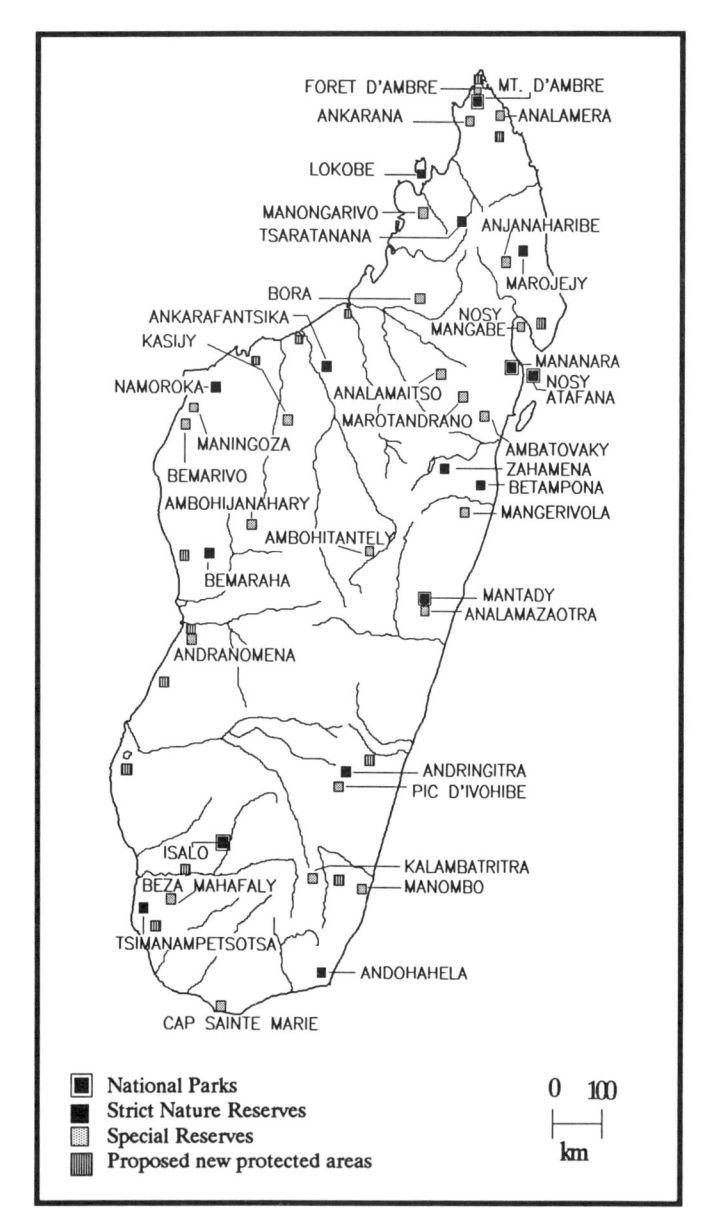

Figure 2.2 Existing and proposed protected areas in Madagascar.

2.4 Species of Conservation Concern

In this section, brief details are given of 18 tenrec species that are of global conservation concern. Most of these are shrew-tenrecs in the genus *Microgale* and, as explained above, further research might indicate that fewer species are involved. The reader is advised to refer to Tables 2.1 and 2.2 while studying these accounts to gain a clearer picture of the taxonomic complexities, and to Fig. 2.2, a map of existing and proposed protected areas in Madagascar, which indicates the location of several places mentioned in the text.

Aquatic Tenrec (*Limnogale mergulus*)

IUCN Category of Threat: Indeterminate.

Distribution: This remarkable species has rarely been recorded, and its distribution is not fully known. Malzy (1965) obtained several specimens from the Ampansandrano Forestry Station on the Ankaratra Massif 35 km north-west of Antsirabe. Gould and Eisenberg (1966) reported signs of the species to be abundant during their visit but did not capture or see animals. M.E. Nicoll and P.J. Stephenson (pers. obs.) visited the site in 1988 and saw only a few scattered signs. Possible feeding signs also suggest the tenrec's presence in a small river 15 km north of Antanifotsy village 35 km south of Ambalavao. Local villagers claimed that the animal occurs there, but no faeces were found. It is not known in similar rivers 20 km to the south within Andringitra Strict Nature Reserve. Gould and Eisenberg (1966) reported the animal 10 km north of Andekaleka (Rogez), but it was apparently rare in this area. The forests at this locality have now been cleared, and no recent information on the animal is available. B.K. Creighton (pers. comm.) found an individual acidentally drowned in an eel trap at Ranomafana Est, 60 km east of Fianarantsoa in August 1988, constituting a new locality record. Five specimens have since been catured alive at this site (R. David Stone, E. Gould, pers. comm.). M.E. Nicoll and P.J. Stephenson (pers. obs.) located a new site 35 km south of Antsirabe where they saw one individual and found abundant faeces on rocks in a small river. Villagers report that the aquatic tenrec occurs in scattered localities on the eastern edge of the central plateau between Antsirabe and Fianarantsoa, but indicate highly localised populations. Villagers always consider that this species is rare. A summary of the currently known distribution of the species is provided in Fig. 2.3.

Ecology and Habits: This semi-aquatic tenrec occurs along small streams and rivers in the east from 450-2,000 m altitude. Gould and Eisenberg (1966) and Eisenberg and Gould (1970) suggest that this species favours water courses where the aquatic plants *Aponogeton* spp. and *Hydrostachis* spp. occur, providing abundant sites for invertebrate prey. Recent observations (M.E. Nicoll and P.J. Stephenson, pers. obs.) show that

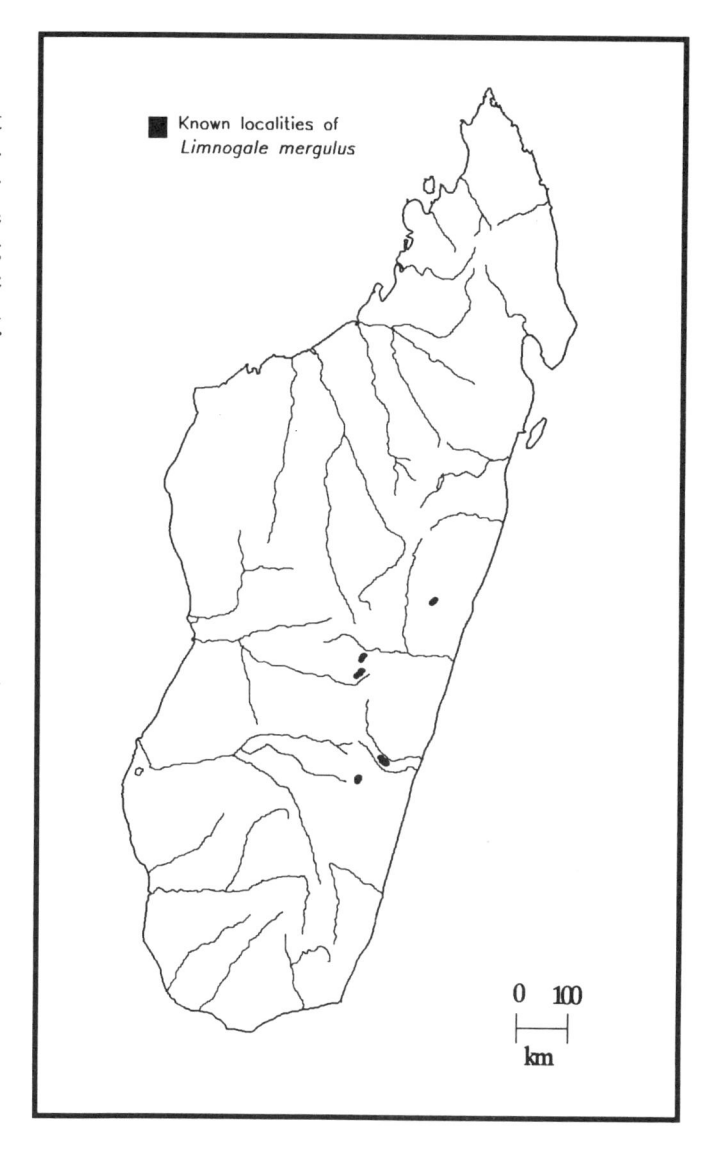

Figure 2.3 Known distribution of the aquatic tenrec (*Limnogale mergulus*), in Madagascar. This little-known species favours fast-flowing streams in mountain areas.

these plants need not be present nor does their occurrence ensure the presence of the tenrec. The main habitat requirement seems to be permanent, clean, fast-flowing water, though many such localities support no populations.

Captivity Records: One adult male was held at the Parc Botanique et Zoologique de Tsimbazaza, Antananarivo, for three weeks in June 1989 before re-release.

Note: The aquatic tenrec has been a major focus of attention for the Insectivore, Tree-Shrew and Elephant-Shrew Specialist Group. Fieldwork has started, and details are provided in the project proposal which is included here as Appendix 1.

Large-eared Tenrec (*Geogale aurita*)

IUCN Category of Threat: *G.a. aurita* - Insufficiently Known; *G.a. orientalis* - Indeterminate.

Distribution: *G.a. aurita* has been widely collected widely in the south-west and west, as far north as the Tsiribihina River. Precise distribution limits are unknown. Recent records indicate that it is abundant at Beza Mahafaly Special Reserve in the south-west and at Morondava in the centre-west, and that it also occurs in the Zombitse Forest near Sakaraha. It has recently been collected in eastern rainforests in the extreme south-east (B.K. Creighton, pers. comm.). *G.a. orientalis* is known from a single specimen captured at Fenoarivo on the north-east coast.

The large-eared tenrec (*Geogale aurita*)
(Photo by M. E. Nicoll/BIOS)

Ecology and Habits: Information is available only for *G. a. aurita*. Gould and Eisenberg (1966) reported one live individual near Morondava which was found torpid under a fallen log in June. M.E. Nicoll and F. Rokotodraparany found numerous individuals within fallen logs in the south-west in May. All were torpid by day but active by night. This species often shares fallen wood with invertebrates, and takes a wide range of invertebrate prey, while showing a marked preference for termites. It is nocturnal. It has been recorded in spiny thorn bush, deciduous forest, and gallery forest dominated by *Tamerindus indica*. Six births of 1-5 offspring in November-February have been recorded (F. Rakotodraparany, pers. comm., P.J. Stephenson, pers.comm.).

Captivity Records: In 1985-1987, 1-6 individuals were maintained in the Parc Botanique et Zoologique de Tsimbazaza in Antananarivo. A female gave birth to one offspring in January 1987, but both subsequently died. Parc Tsimbazaza has more recently maintained a population of more than 10 adults, among which five females have given birth. Longevity in captivity of a male caught as an adult was 25 months.

Short-tailed Shrew-tenrec (*Microgale brevicaudata*)

IUCN Category of Threat: Insufficiently Known.

Distribution: Found in the village of Antsirabe Avaratra (13°58'S, 49°58'E), Antalaha, Antsiranana Province (MacPhee 1987), and Manongarivo Special Reserve in north-western Sambirano rainforests, Ambanja, Antsiranana Province. MacPhee (1987) considers *M. breviceps* and *M. (Paramicrogale) occidentalis* to belong to this species.

Western Short-tailed Shrew-tenrec (*Microgale (Paramicrogale) occidentalis*)

IUCN Category of Threat: Indeterminate.

Distribution: The type specimen was captured near Andriafavelo village, at 80 m altitude and 20 km from sea, 42 km northeast of Maintirano on the north-west coast (MacPhee 1987). A possible second specimen was collected, but not verified by museum collections. In the collection area only small remnants of western dry forest remain, otherwise having been replaced with man-induced grasslands. There are no records since the specimen was described by Grandidier and Petit (1931), but three specimens that fit this species' description have been captured at Morondava further south. Although it is officially Extinct according to the IUCN categories of threat, as it has not been seen in the last 50 years, it is given Indeterminate status at present because of lack of recent collecting effort in Maintirano area. MacPhee (1987) considers this form to be the same species as *M. brevicaudata*.

Large-footed Shrew-tenrec or Cowan's Shrew-tenrec (*Microgale crassipes*)

IUCN Category of Threat: Insufficiently Known.

Distribution: A single specimen was collected near Antananarivo (18°55'S, 47°32'E), but the precise locality is unknown. It is considered by MacPhee (1987) to be the same species as the widespread *M. cowani* which occurs widely in central plateau forest remnants. *M. cowani* is variable in its characteristics and MacPhee's diagnosis is probably correct, with only the ear length being noticeably shorter. Although it has not been seen in the last 50 years and therefore it is Extinct by IUCN definitions, it is retained as Insufficiently Known in light of the taxonomic difficulties.

Drouhard's Shrew-tenrec or Cowan's Shrew-tenrec
(*Microgale drouhardi*)

IUCN Category of Threat: Insufficiently Known.

Distribution: Known from seven specimens from the region of Antsiranana (12°16'S, 49°18'E) in the extreme north. Exact locality data are missing, which is unfortunate owing to the wide diversity of ecological conditions in the area. The extinct volcanic massif, Ambohitra (Montagne d'Ambre), has on its higher slopes humid rainforests that are floristically similar to eastern forests, while the lower slopes support either man-induced grassland or drier deciduous forest typical of the west. The neighbouring Tendrombohitr'Antsigy (Montagne des Français), Ankarana, Analamera and Cap d'Ambre areas support different dry forest forms on limestone formations. MacPhee (1987) regards this species as being the same species as *M. cowani*, but although the *drouhardi* specimens are reportedly juvenile they are consistently larger than *cowani* and have a relatively longer tail and hind foot. We consider it more likely that *M. drouhardi* is the same species as *M. taiva*. It is Extinct by IUCN definitions, but is retained as Insufficiently Known owing to lack of recent collecting effort, isolation from other known *cowani*-like tenrecs, and taxonomic confusion.

Long-nosed Shrew-tenrec or Cowan's Shrew-tenrec
(*Microgale longirostris*)

IUCN Category of Threat: Insufficiently Known.

Distribution: The type specimen is from Ampitambe (which may be the same as Ampitabe) Forest in north-eastern Betsileo. The precise locality is unknown. Eisenberg and Gould (1970) list this species at Didy, south-east of Lake Aloatra in the eastern rainforests, and at Andasibe. MacPhee (1987) considers this species to be the same species as *M. cowani*, but Eisenberg and Gould (1970) recorded both present at Didy and Andasibe. One specimen thought to be *M. longirostris* was recently caught 120 km east of Antananarivo at Anjozorobe, and is distinctly larger than *cowani* and with a longer snout.

Cowan's shrew-tenrec (*Microgale cowani*)
(Photo by M. E. Nicoll/BIOS)

Striped Shrew-tenrec or Cowan's Shrew-tenrec
(*Microgale melanorrhachis*)

IUCN Category of Threat: Insufficiently Known.

Distribution: Recorded from Andasibe (Périnet) (19°00'S, 48°30'E) in eastern rainforests at 980 m altitude, and from Ambatovaky Special Reserve 200 km further north (C. Raxworthy, pers. comm.). MacPhee (1987) considers this form to be the same species as *M. cowani*, with *M. melanorrachis* being a juvenile form. However, adults of both *M. cowani* and *M. melanorrachis* occur with overlapping ranges at Andasibe, and both should be retained as full species.

Taivi Shrew-tenrec or Cowan's Shrew-tenrec
(*Microgale taiva*)

IUCN Category of Threat: Insufficiently Known.

Distribution: Recorded from Ambohimotombo Forest (20°43'S, 47°26'E) at 1,300 m, south-east of Ambositra (although this is only a best estimate of the precise locality). Specimens thought to be of this species, including a pregnant female, were recently captured at Andasibe (Périnet). MacPhee (1987) regards this as the same species as *M. cowani*, but this is surprising as it would form a distinctive isolate within the geographical range of otherwise 'typical' series of *cowani*, and it has a consistently longer head-body length, with relatively greater tail and hind-foot length, and greater weight, similar to *drouhardi*, with which it is probably conspecific. It is Extinct according to IUCN definitions, but is retained as Insufficiently Known because of only infrequent recent collecting within its range, and its confused taxonomic status.

Lesser Long-tailed Shrew-tenrec
(*Microgale longicaudata*)

IUCN Category of Threat: Insufficiently Known.

Distribution: Specimens were collected at Ankafina (21°12'S, 47°13'E), a 1,600 m high hill south of Ambohimahasoa, and at the extreme present western limit of the eastern rainforests. Recent specimens thought to be of this species have been recorded from Ambatovaky Special Reserve in the north-east (C. Raxworthy, pers. comm.). MacPhee (1987) considers *M. prolixacaudata* to be conspecific, but *prolixacaudata* is recorded at the furthest distance possible from *longicaudata*, in the extreme north, and the specimen was juvenile. It is Extinct according to IUCN definitions, but is retained as Insufficiently Known owing to lack of recent collecting within its range.

Major's Lesser Long-tailed Shrew-tenrec (*Microgale majori*)

IUCN Category of Threat: Insufficiently Known.

Distribution: The only specimens are from the same locality as *M. longicaudata*. They are distinguished from *M. longicaudata* by their shorter tail, but the single entire specimen apparently had an aberrant tail. MacPhee (1987) considers this species to be the same species as *M. longicaudata*. It is officially Extinct according to IUCN definitions, but is retained as Insufficiently Known owing to lack of recent collecting within its geographical range, and taxonomic uncertainty.

Northern Lesser Long-tailed Shrew-tenrec (*Microgale prolixacaudata*)

IUCN Category of Threat: Indeterminate.

Distribution: Known from a single juvenile specimen from the Antsiranana region, the same localities as *drouhardi*. It is considered to be the same species as *longicaudata* by MacPhee (1987). It is Extinct according to IUCN definitions, but is retained as Indeterminate owing to lack of collecting within its geographical range, and taxonomic uncertainty.

Pygmy Shrew-tenrec (*Microgale parvula*)

IUCN Category of Threat: Insufficiently Known.

Distribution: A single immature specimen was collected in the same locality as *M. drouhardi*. It is similar to *M. pulla*, collected in 1986 by P.J. Stephenson at Andrivola Forest (15°46'S, 49°35'E), about 40 km south-west of Maroantsetra in the north-east. Jenkins (1988) notes the similarities between the two species, but considers *pulla* not to be simply an adult *parvula*, but a full species. *M. parvula* is Extinct according to IUCN definitions, but is retained as Insufficiently Known owing to lack of recent collecting in its geographical range.

Greater Long-tailed Shrew-tenrec (*Microgale principula*)

IUCN Category of Threat: Insufficiently Known.

Distribution: Known from a single specimen from Midongy Atsimo (Midongy du Sud) (23°35'S, 47°01'E) in the south-eastern rainforests at 500 m altitude. MacPhee (1987) considers *M. sorella* which is found further north to be the same species. It is Extinct according to IUCN definitions, but is retained as Insufficiently Known owing to lack of recent collecting within its geographical range, and taxonomic uncertainty.

Long-tailed Shrew-tenrec or Greater Long-tailed Shrew-tenrec (*Microgale sorella*)

IUCN Category of Threat: Insufficiently Known.

Distribution: The single adult male specimen is from Beforona village (18°48'S, 48°35'E), at 500 m altitude in rainforest on the road from Antananarivo to Toamasina. MacPhee (1987) considers *sorella* to be the same species as *principula*. It is Extinct, but is retained as Indeterminate owing to lack of recent collecting in its geographical range, and taxonomic uncertainty.

Thomas's Shrew-tenrec (*Microgale thomasi*)

IUCN Category of Threat: Insufficiently Known.

Distribution: The type specimen is from the same locality as *M. taiva*. Recent specimens captured by M.E. Nicoll, B.K Creighton, and P.J. Stephenson at Ranomafana Est, east of Fianarantsoa, and at Anjazorobe at altitudes of 800 - 1,200 m in eastern rainforests have been assigned to *M. thomasi*.

Gracile Shrew-tenrec (*Microgale gracilis*)

IUCN Category of Threat: Insufficiently Known.

Distribution: The type specimen was collected at the same locality as the type specimen of *M. taiva*. Additional specimens are now known from Andringitra Strict Nature Reserve, Ranomafana Est, and Andasibe (Périnet). Few individuals have been recorded.

Dark Pygmy Shrew-tenrec (*Microgale pulla*)

IUCN Category of Threat: Insufficiently Known.

Distribution: A single specimen of this recently described species was captured in Andrivola Forest (15°46'S, 49°36'E), about 40 km south-west of Maroantsetra in the north-eastern rainforest. It is possible that *pulla* is an adult form of *parvula*, but Jenkins (1988) considers that it is a distinct species.

Chapter 3: The Insectivora of Africa

3.1 Introduction

Africa and its in-shore islands contain a rich assemblage of the Insectivora, with four distinct lineages represented. Two of these, the golden moles (family Chrysochloridae) and the otter-shrews (family Tenrecidae) comprise the living members of ancient mammalian radiations, and are confined to the continent. The African shrews (Soricidae) are exceptionally rich in diversity, with a high proportion being endemic to the continent, and the hedgehogs (Erinaceidae) are well represented.

Producing an Action Plan for the African Insectivora presents some difficulties, particularly on account of the scarcity of information on distribution and abundance, and taxonomic uncertainties in three of the families present. Consequently, considerable emphasis has to be placed on increasing our knowledge of African Insectivora in order to help resolve the above problems.

Assessment of African Insectivora conservation needs has only been carried out in South Africa (Meester 1976; Smithers 1986), and even in this well-studied country there remain numerous distributional and taxonomic problems to resolve. Elsewhere, the most complete information comes from the central African countries of Rwanda and Burundi (Hutterer *et al.* 1987), or for individual threatened species, such as the Nimba otter-shrew (*Microgale lamottei*) (Vogel 1983). The great majority of the African Insectivora remain virtually unknown, and even for the species that are known to be threatened, there is often insufficient information to determine the level of risk involved or to propose potential solutions. It is, however, reasonable to assume that the majority of the species can be protected for the future if adequate areas of habitat can be conserved.

A principal task of this Action Plan is thus to bring together and update available information into one document, and to identify other species that might also be threatened. In addition, the Action Plan attempts to identify sites where the Insectivora are at risk through habitat loss in centres of species richness or high local endemism. The latter may reinforce attempts at conservation lobbying for areas of general biological importance, and combined with the individual species treatments, should provide guidance for future Insectivora conservation programmes for the entire continent.

Owing to the inherent difficulties in dealing with the African Insectivora, this section first gives a brief overview of classification, followed by a description of ecosystem diversity in Africa. This is followed by a taxonomic review of individual species potentially at risk.

3.2 Classification

The Insectivora comprise a series of taxonomic groups that in reality may form long-separated and distinctive lineages, but are placed in the same order because they share a range of primitive characteristics. Concerning Afro-Malagasy species, Van Valen (1967) considered the Erinaceidae and Soricidae to be the only families within the Insectivora, separating the Chrysochloridae and Tenrecidae into a separate order, the Deltatheridea. McKenna (1975) distinguished the Erinaceidae as a different order from the Chrysochloridae, Soricidae and Tenrecidae. There is also some dispute over classification at the generic and species level.

This Action Plan retains the four African families within the single order Insectivora. These families are: the hedgehogs (Erinaceidae: three genera, five species); shrews (Soricidae: seven genera, about 136 species); golden moles (Chrysochloridae: seven genera, 18 species); and otter-shrews (Tenrecidae: two genera, three species) (see Table 3.1). The names of species and genera used here are those listed in Meester and Setzer (1974), or in Hutterer *et al.* (1987), from which any later revisions may be readily traced.

Table 3.1: Taxonomic summary of the African Insectivora (excluding Madagascar)

Hedgehogs: Family Erinaceidae

Genus	*Atelerix*	3 species
	Paraechinus	1 species
	Hemiechinus	1 species

Golden Moles: Family Chrysochloridae

Genus	*Cryptochloris*	2 species
	Chrysospalax	2 species
	Chrysochloris	3 species
	Eremitalpa	1 species
	Chlorotalpa	5 species
	Calcochloris	1 species
	Amblysomus	4 species

Otter-shrews: Family Tenrecidae

Genus	*Potamogale*	1 species
	Micropotamogale	2 species

Shrews: Family Soricidae

Genus	*Crocidura*	104 species
	Paracrocidura	3 species
	Ruwenzorisex	1 species
	Suncus	6 species
	Sylvisorex	8 species
	Myosorex	13 species
	Scutisorex	1 species

The golden moles are restricted to Africa, with all species confined to the sub-Saharan region and most occurring in the southern part of the continent. The two African otter-shrew genera are restricted to the rainforests and their fringes in western and central Africa, the remaining eight genera of tenrecs being confined to Madagascar. The hedgehogs are widespread in Africa and Eurasia. No genera are endemic to Africa, but two species within the genus *Atelerix* are confined to this continent. Within the shrews, all but two of the 104 *Crocidura* species are restricted to Africa and its inshore islands, four of the five native *Suncus* species are endemic to Africa, and the genera *Sylvisorex* (eight species), *Paracrocidura* (three species), *Myosorex* (13 species), *Ruwenzorisorex* (one species), and *Scutisorex* (one species) are endemic to the continent.

3.3 Biogeographic Regions of Africa

The Insectivora occur widely in Africa and its inshore islands, inhabiting a broad range of distinctive ecosystems. White (1983) provides a recent review of phytogeographic zones, describing representative ecosystems and providing maps. In broad terms, the vegetation zones of continental Africa can be considered to comprise montane (both forest and savanna), lowland forest, moist savanna, woodland, dry savanna with bush and thicket, and desert and semi-desert. The inshore islands considered here support a natural vegetation of moist forest.

The hedgehogs occur in all environments except dense moist forest. Among the shrews, the genera *Sylvisorex*, *Myosorex*, *Paracrocidura*, *Ruwenzorisorex* and *Scutisorex* are largely restricted to moist forest or woodland environments. Many *Crocidura* species show the same preference, but this genus also occurs in dry savanna or semi-desert conditions. The golden moles occur in deserts, woodlands, and moist forests, and two of the otter-shrews are restricted to moist forests and their fringes.

A review of Insectivora biogeographic patterns within Africa is constrained by unequal collecting effort in different countries and uncertainty in taxonomic validity. Small mammals have not been subjected to the same level of study as larger mammals or birds, and among the smaller terrestrial species, research has focused on rodents. However, reviews of Insectivora distribution in central (Hutterer *et al.* 1987), eastern (Kingdon 1974), and southern Africa (Meester 1976; Smithers 1986) provide some useful conclusions concerning biogeographic patterns and conservation requirements.

Species richness, distribution patterns, and the degree of local or regional specific endemism differ between the families of African Insectivora.

The hedgehogs, although not diverse in Africa, reach their greatest diversity in the northern part of the continent where two or three species may occur with overlapping ranges. This is related in part to the ranges of Eurasian species extending into adjacent regions of Africa. In sub-Saharan Africa, only one species is normally present in any given locality.

The diversity of African shrews tends to be greatest in countries with large surface areas and moist forest habitats, and least in arid countries (see Table 1.1; Hutterer *et al.* 1987). Greatest species richness occurs in two forest countries, Zaïre (by far the highest), and Cameroon. Species richness is also high in some countries with mixed forest and savanna environments, and high topographical variation, notably in Kenya, Tanzania, Uganda, Nigeria, Sudan and Ethiopia. The number of species in Rwanda (15) is remarkably high in view of the tiny size of the country. These findings indicate that shrew diversity corresponds closely with known Pleistocene refugia and centres of endemism for other groups of vertebrates (Rodgers *et al.* 1982; Prigogine 1985; Stuart 1985). The richest refugia are centred in two mountain areas, the western Cameroon highlands (extending into eastern Nigeria), and the Albertine Rift mountains in eastern Zaïre, western Uganda and Rwanda, and are associated with areas of moist montane forest. Other refugia areas that are not quite as rich in endemic species, but which are nevertheless important, include the Ethiopian highlands, the Nimba massif in Liberia, Guinea and Ivory Coast, the mountains of eastern Tanzania, and the Kenyan coastal forests. Hutterer *et al.* (1987) noted that the Albertine Rift forests support two distinct types of shrew populations. Widely distributed species occur in this region, but are mostly in habitats such as *Eucalyptus* forest, moist savannas, and cultivated lands. The second group of seven species, which he termed 'central African endemics', are entirely restricted to natural montane forest remnants or their edges. Hutterer and Happold (1983) have demonstrated that forest and savanna shrew species show a similar separation in Nigeria. Further research is needed to establish whether similar patterns of ecological separation occur in other refugia areas.

Golden moles occur in habitats ranging from desert to moist forest, but differ from other Insectivora families in that they have a marked southern African centre of diversity. Only three of the 18 species occur outside southern Africa, in central Africa and Somalia (there being just one specimen of the Somali golden mole (*Chlorotalpa tytonis*). The southern species themselves vary from being widespread within the sub-region, with a rather catholic use of habitats, to having very restricted ranges that are generally confined to a particular series of soil or habitat conditions.

The largest African member of the otter-shrews is restricted to the central moist forest block and its fringes. This species, the giant otter-shrew (*Potamogale velox*), faces no threat of extinction as long as forest occurs adjacent to suitable water courses, even in areas where forest has been cleared extensively on surrounding elevated ground (Nicoll 1985). The two lesser otter-shrews (*Micropotamogale*) are restricted to two montane refugia: the Ruwenzori otter-shrew (*M. ruwenzorii*) is found only in the mountains and surrounding areas of north-eastern Zaïre and western Uganda (Rahm 1966; Kuhn 1971; Kingdon 1974); and the Nimba otter-shrew (*M. lamottei*) is found only on Mount Nimba (Vogel 1983). The Nimba otter-shrew resembles the giant otter-shrew in requiring moist forest habitats whereas the Ruwenzori otter-shrew survives in a wide range of secondary habitats.

3.4 Species of Conservation Concern

In this section, details are given of one species of hedgehog, 13 golden moles, two otter-shrews, and 42 shrews that are of conservation concern. Less information is available on the shrews, in comparison with the other families, and this shows up in the species accounts below.

Family Erinaceidae: the Hedgehogs

Three genera and five species occur in Africa, although Wozencraft (pers. comm.) considers that *Paraechinus* is a subgenus of *Hemichinus*. Only one species, the South African hedgehog (*Atelerix frontalis*), has been accorded an IUCN Threatened Species status. The four-toed hedgehog (*A. albiventris*), is collected for sale to private collectors and zoological gardens, but there is no evidence that the species requires special protection beyond any existing national legal protection. The other species, the Algerian hedgehog (*A. algirus*), desert hedgehog (*Paraechinus ethiopicus*), and long-eared hedgehog (*Hemiechinus auritus*) all have reasonably wide distributions, and there is no indication of threat or the need for conservation action.

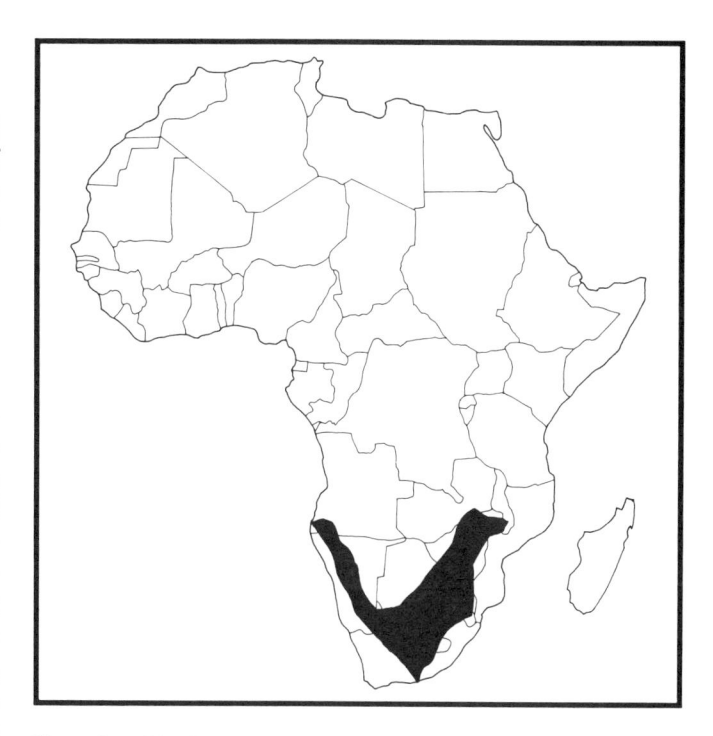

Figure 3.1   Distribution of the South African hedgehog (*Atelerix frontalis*)

South African Hedgehog (*Atelerix frontalis*)

IUCN Category of Threat: Rare

This species occurs in two discrete areas in southern Africa. One population occupies an eastern range, encompassing South Africa, Lesotho, Botswana, Zimbabwe and Mozambique, while a western population occurs in Angola and Namibia (Smithers 1986). There is insufficient information to indicate whether the species occurs in localities between these two populations. Its distribution is indicated in Fig. 3.1.

Smithers (1986) reports that this species occurs in thornveld and karroid formations, and has adapted to garden environments in urban and suburban localities in South Africa. There is insufficient information to determine whether this species is declining, but individuals are frequently killed on roads, hunted for food or collected as pets, and they do not adapt well to agricultural development. It occurs in several reserves, and the Bloemfontein Zoo possessed two males and five females in 1986 (Smithers 1986).

The South African hedgehog is protected under Schedule 2 in Cape Province and Transvaal, and under Schedule 1 in the Orange Free State, and may be kept in captivity only under permits. Its presence in Natal is not confirmed. Smithers (1986) recommends that research be carried out on this species' ecology, and that there should be a strict enforcement of existing protective legislation.

Family Chrysochloridae: the Golden Moles

The golden moles are represented by seven genera and 18 species, all restricted to sub-Saharan Africa. Thirteen of these species are given IUCN Threatened Species Categories here, with two designated as Vulnerable, two Rare, and nine Indeterminate. The distribution of all 18 species is indicated in Figure 3.2.

Hottentot golden mole (*Amblysomus hottentotus*)
(Photo by G. C. Hickman)

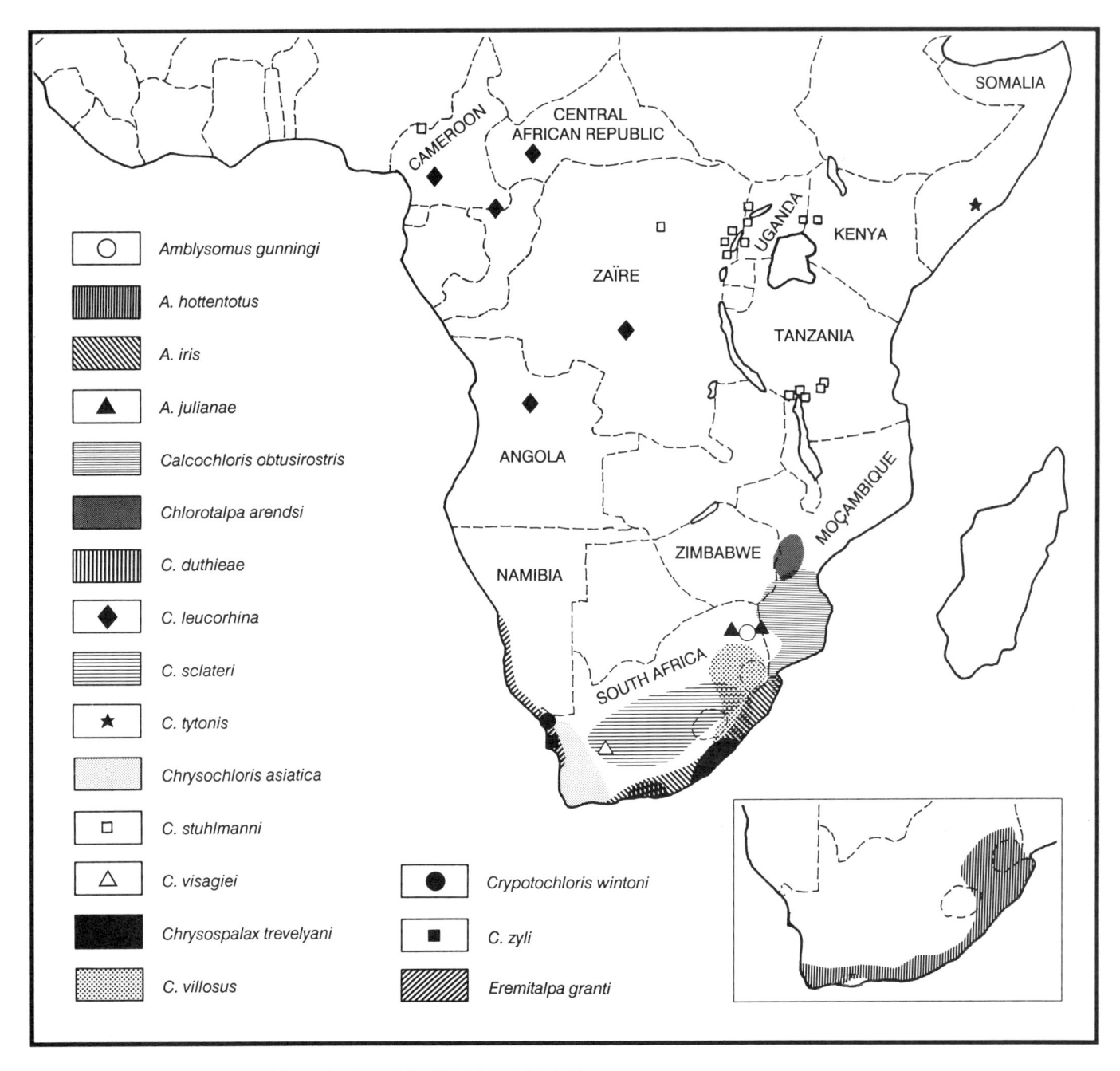

Figure 3.2 Distribution of the golden moles (copyright: Wiley-Liss Publ., N.Y.)

Of the non-threatened species, the Cape golden mole (*Chrysochloris asiatica*) occurs in western Cape Province, South Africa. It has adapted well to man-made environments including gardens. There is no indication of a population decline or need for threatened species status. Stuhlmann's golden mole (*Chrysochloris stuhlmanni*) occurs in northern Zaïre, Uganda, and Tanzania. In Uganda it occurs up to an altitude of 2,800 m. It is well represented in museum collections and occurs in protected areas. For Arends' golden mole (*Chlorotalpa arendsi*), which occurs in eastern Zimbabwe; the Congo golden mole (*Chlorotalpa leucorhina*), which occurs in Cameroon, southern Zaïre, and northern Angola; and the Hottentot golden mole (*Amblysomus hottentotus*), which occurs in South Africa, there are no indications of population declines or of needs for threatened species categories. A general proposal for golden mole conservation is presented in Appendix 2.

De Winton's Golden Mole (*Cryptochloris wintoni*)

IUCN Category of Threat: Indeterminate

This species is known from specimens collected in sand dunes at Port Nolloth in Cape Province, South Africa (Smithers 1986). Meester (1976) originally assigned an IUCN status of Rare to this species, but Smithers re-designated this to Indeterminate pending the availability of further information.

Van Zyl's Golden Mole (*Cryptochloris zyli*)

IUCN Category of Threat: Indeterminate

The type specimen is from Lambert's Bay in Cape Province, South Africa. Several other museum specimens exist. Meester (1976) originally assigned an IUCN status of Rare to this species, but Smithers re-designated this to Indeterminate pending the availability of further information.

Rough-haired Golden Mole (*Chrysospalax villosus*)

IUCN Category of Threat: Vulnerable

This species occurs in Transvaal, Natal, and eastern Cape Province in South Africa (Meester and Setzer 1974; Nowak and Paradiso 1983). The rough-haired golden mole occurs in dry grassy habitats, particularly in areas bordering marshes. In Transvaal they are uncommon (Smithers 1986), there is no known area where they can be reliably trapped, and only two specimens have been collected in the last 15 years (G. Hickman, pers. comm.). Without further information it is difficult to assess what conservation measures might be necessary.

Giant Golden Mole (*Chrysospalax trevelyani*)

IUCN Category of Threat: Rare

The range of this species is limited to a few small areas in eastern Cape Province from the King William's Town and East London districts eastwards to Port St. Johns in Transkei, and marginally into Ciskei, South Africa. The giant golden mole occurs in forests and valley forests where there are deep soils, leaf litter, and dense shrubs. These habitats are being cleared and degraded, especially where they are close to human settlements. Clearance and habitat damage occurs largely because of firewood collection, barkstripping, cutting for construction, and livestock ranging in the forest. Dogs prey upon giant golden moles. Only one locality in the range of this species is fenced, thereby providing protection from dogs.

Smithers (1986) suggests that rigid control of remaining habitats should be implemented, and that a reserve be created to protect a representative sample of the otherwise rapidly degrading habitat. It is also suggested that this species should be protected under Cape Province legislation.

Visagie's Golden Mole (*Chrysochloris visagiei*)

IUCN Category of Threat: Indeterminate

Visagie's golden mole is known only from a single specimen collected at Gouna in Cape Province, South Africa. Meester and Setzer (1974) suggest that this specimen may simply represent an aberrant form of the Cape golden mole, and its taxonomic status remains in doubt. Meester (1976) suggested a threatened species category of Rare, but Smithers (1986) reclassified it as Indeterminate pending the availability of further information.

Grant's Golden Mole (*Eremitalpa granti*)

IUCN Category of Threat: Rare

This species occurs in a narrow belt in south-western South Africa and in the Namib Desert of Namibia. It occurs on coastal sand dunes and also inland, showing a preference for areas with scattered clumps of the dune grass, *Aristidia sabulicola*, at least in South Africa (Coetzee 1969). It does not occur in any protected area in South Africa, but if the proposed Groen River National Park is established, it will cover part of the species' range. The Namib Desert National Park in Namibia is within the species' range.

Duthie's Golden Mole (*Chlorotalpa duthiae*)

IUCN Category of Threat: Indeterminate

Duthie's golden mole occurs in a narrow coastal band between Knysna and Port Elizabeth in southern Cape Province in South Africa. The habitat is typified by alluvial sands and sandy loams. It may be locally common, but little is known of its ecology, and few museum specimens exist (Smithers 1986).

Sclater's Golden Mole (*Chlorotalpa sclateri*)

IUCN Category of Threat: Indeterminate

This species occurs in a series of scattered localities from Cape Province north eastwards to south-east Transvaal, with additional sites from eastern Orange Free State and Lesotho. Virtually nothing is known about the species, and its status is difficult to determine (Smithers 1986).

Somali Golden Mole (*Chlorotalpa tytonis*)

IUCN Category of Threat: Indeterminate

This species is known only from a single specimen collected at Giohar in Somalia. Given the limited information and its unusual locality, a threatened species category of Indeterminate is appropriate.

Yellow Golden Mole (*Chlorotalpa obtusirostris*)

IUCN Category of Threat: Rare

The yellow golden mole occurs in southern Zimbabwe, southern Mozambique, and north-eastern South Africa. In South Africa this species favours alluvial soils, dry river beds, and coastal sand dunes. It occurs in the Kruger National Park. Little is known of its habits and ecology.

Gunning's Golden Mole (*Amblysomus gunningi*)

IUCN Category of Threat: Indeterminate

Gunning's golden mole is known only from the Woodbrush Forest, and the Agatha Forest 20 km to the south, in north-eastern Transvaal (Rautenbach 1978). It appears to be associated with montane forest, but has also been captured in montane grasslands and ploughed land. Its range is apparently very restricted, and only seven museum specimens are known (Smithers 1986).

Zulu Golden Mole (*Amblysomus iris*)

IUCN Category of Threat: Indeterminate

This species occurs widely from Knysna in Cape Province to Natal, South Africa, but is restricted to a narrow coastal band. It appears to be associated with light sandy soils (Smithers 1986).

Juliana's Golden Mole (*Amblysomus julianae*)

IUCN Category of Threat: Indeterminate

Only five museum specimens are known from three sites near Pretoria and two in the Kruger National Park, in South Africa. Thus insufficient evidence is available to determine its current status, and a category of Indeterminate is appropriate.

Family Tenrecidae: the Otter-shrews

Three species in two genera of the Tenrecidae occur in sub-Saharan Africa, restricted to the western and central rainforest blocks, and their margins. The majority of tenrec species are confined to Madagascar, where there are eight genera. There are currently no allocations of threatened species categories for African tenrecids, but two are proposed here. The third species, the giant otter-shrew (*Potamogale velox*) is more widespread, occurring in the central rainforest zone and peripheral areas from Nigeria eastwards to western Kenya, and southwards to central Angola and northern Zambia. It occurs from sea level up to 1,800 m, and occurs in slow-flowing streams, forest pools, and montane torrents. In Cameroon, forest clearance and subsequent soil erosion with increased opaqueness of water courses is leading to local disappearance of this species (Nicoll 1985). The species is widely hunted for its skin and is also trapped accidentally, but if the forest habitat remains relatively intact, even as a narrow strip along river banks, the species appears to maintain viable populations. However, it would be prudent to keep the status of the species under review, especially in the event of major perturbations in the rainforest ecosystems.

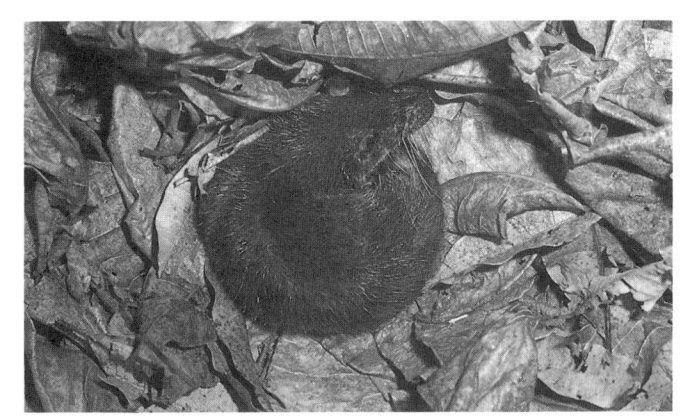

The giant otter-shrew (*Potamogale velox*)
(Photo by M. E. Nicoll/BIOS)

Nimba Otter-shrew (*Micropotamogale lamottei*)

IUCN Category of Threat: Endangered

The Nimba otter-shrew is restricted to the Mount Nimba area spanning the borders of Ivory Coast, Liberia and Guinea (see Fig. 3.3). Almost all known specimens of this species, which was first described in 1954 (Heim de Balsac 1954), have been captured in an area covering less than 1,500 km². Surveys carried out by Vogel (1983) indicate that this species is relatively common in the Danané-Man region of Ivory Coast, and that similar populations occur in the same habitats in neigh-

The Nimba otter-shrew (*Micropotamogale lamottei*)
(Photo by P. Vogel)

bouring Guinea and Liberia. It occurs in swampy areas and in small rivers and forest streams. Vogel (1983) successfully maintained animals in captivity for some months and reared young individuals.

The long-term future of this species seems bleak. Mining activities have devastated the Liberian sector of the mountain and habitat conservation is generally ineffective. Reports of the potential expansion of bauxite mining to Guinea are alarming, and any such developments should be strongly opposed. The current World Bank programme of assistance to form a Biosphere Reserve on the Guinean side deserves strong support. Further research and perhaps a captive breeding programme are recommended as urgent priorities.

Ruwenzori Otter-shrew (*Micropotamogale ruwenzorii*)

IUCN Category of Threat: Indeterminate

This species was first discovered in 1953 and, since then, relatively few specimens have been captured, most being found in fish-traps (Kingdon 1974). To date, all specimens have come from the Kivu and Ruwenzori regions of eastern Zaïre and western Uganda (Rahm 1966) (see Fig. 3.3). Vegetation and altitude appear to have little influence on distribution, as it has been recorded in montane forest, lowland forest, savanna, and cultivated regions. Its restricted distribution suggests it is a relictual population associated with the Albertine Rift refugium.

This species is designated as Indeterminate pending further information. Further research and a possible captive propagation programme are strongly recommended.

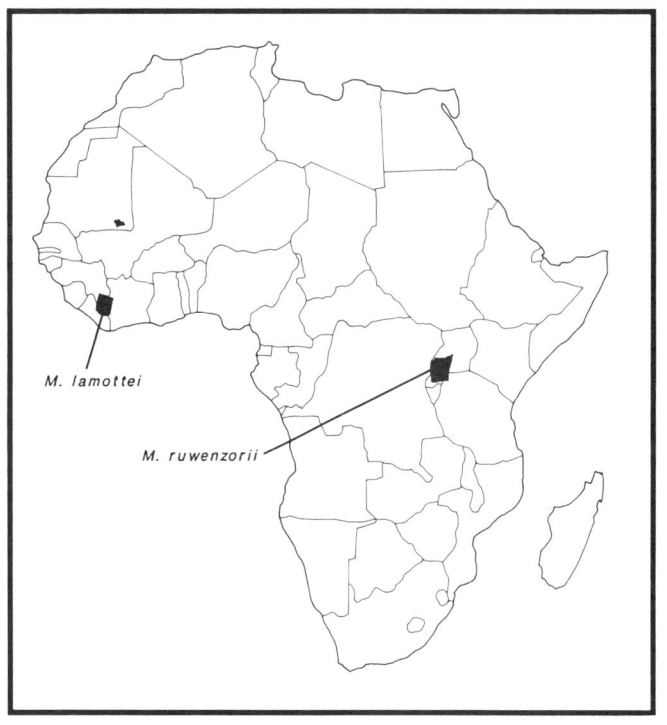

Figure 3.3 Distribution of two African tenrecs, the Nimba otter-shrew (*Micropotamogale lamottei*) and the Ruwenzori otter-shrew (*Micropotamogale ruwenzorii*).

Family Soricidae: the Shrews

The African shrews are represented by seven genera and 136 species, of which one species, the musk shrew *Suncus murinus*, is introduced. All are members of the Crocidurinae. Detailed treatment of each species is neither practical or justified given currently available data, and a regional approach to shrew conservation needs, as is done in Chapter 5, is perhaps more valid. However, where possible, selected species are discussed briefly below on an individual basis.

Genus *Crocidura*

Of the 143 *Crocidura* species listed by Nowak and Paradiso (1983), plus several new species subsequently recorded (Hutterer *et al.* 1987), 104 occur on the African continent and its inshore islands, and only two of these species also occur outside the region. The geographical ranges of individual species vary from being widespread to very restricted (as is the case with many of the montane endemics). Because the *Crocidura* are poorly known, only brief notes are provided below on 30 species which are of conservation concern, in view of their limited distributions or local occurrence, often in habitat - such as forests - that is being destroyed. Other species may be similarly threatened but insufficient information is available for an accurate assessment.

Crocidura ansellorum - This species is known from only two males and a female from gallery forest at Kasombu Stream (Mwinilunga District) and Nyansowe Stream (Solwezi District) in north-western Zambia (Hutterer and Dippenaar 1987).

Crocidura baileyi - Endemic to Simien National Park in the Ethiopian highlands, and possibly other localities in Ethiopia west of the Rift Valley.

Crocidura congobelgica - The type locality of this species is Lubila, near Bafwasende, in the Ituri Forest, Zaïre. It appears to have a highly restricted distribution.

Crocidura crenata - This species has been recorded from Makokou and Belinga, Gabon, and also from Zaïre.

Crocidura eisentrauti - Endemic to Mount Cameroon, and found between 2,000 and 3,000 m.

Crocidura glassi - A local species from the Ethiopian highlands east of the Rift Valley, the type locality being the Gara Mulata Mountains.

Crocidura grassei - Very local in Gabon (Belinga), and Central African Republic (Boukoto).

Crocidura kivuana - Highly restricted species from west of Lake Kivu in the Kahuzi Mountains, eastern Zaïre (Hutterer et al. 1987).

Crocidura lanosa - Very local to the Kahuzi Mountains in eastern Zaïre and Nyungwe Forest, Rwanda (Hutterer et al. 1987).

Crocidura latona - Very local, from Medje, Avakubi, Tshuapa, Ubembo and Ikela, all in central Zaïre.

Crocidura longipes - Known only from two swamps in western Nigeria, east of Bahindi, near Kainji Lake National Park (Hutterer and Happold 1983).

Crocidura lucina - Collected at 3,000 m in the Ethiopian highlands east of the Rift Valley around Dinshu and Mount Albasso.

Crocidura ludia - Known only from Medje in Zaïre.

Crocidura manengubae - A species endemic to Cameroon, found at 1800 m near Lake Manenguba in the western highlands (Hutterer 1981).

Crocidura maquassiensis - Only six specimens of this very small shrew have been recorded, two each from Transvaal in South Africa, Swaziland, and Zimbabwe (Smithers 1986).

Crocidura monax - A forest species from northern Tanzania (Mount Kilimanjaro and Uluguru Mountains).

Crocidura nimbae - Only found on Mount Nimba (Guinea, Ivory Coast and Liberia) Hutterer *et al.* 1987)

Crocidura phaeura - A very local forest species from Ethiopia, known only from the type locality at the west base of Mount Gwamba, north-east of Allata.

Crocidura polia - Known only from Medje in Zaïre.

Crocidura raineyi - A very rare species from Mount Gargues, Kenya.

Crocidura selina - A very local species from Mabira Forest, Uganda.

Crocidura stenocephala - Found only in the marshes of Mount Kahuzi, west of Lake Kivu, eastern Zaïre.

Crocidura tansaniana - A species from the Usambara Mountains, Tanzania.

Crocidura telfordi - A species from the Uluguru Mountains, Tanzania.

Crocidura thalia - This species is from north-west Bale Province in the Ethiopian highlands.

Crocidura thomensis - Found only on São Tomé, very rare (Heim de Balsac and Hutterer 1982).

Crocidura usambarae - A species of overlapping range with *C. tansaniana*, from the Usambara Mountains, Tanzania.

Crocidura wimmeri - Known only from moist savanna in Ivory Coast around Adiopodoume, where it is very rare.

Crocidura n.sp. - A forest endemic from Harenna Forest in the Bale Mountains National Park, Ethiopia (not yet described).

Crocidura n.sp. - A second forest endemic from Harenna Forest in the Bale Mountains National Park, Ethiopia (also not yet described).

Crocidura n.sp. - A new species from remnant forests in the Uzungwa Mountains and on Mount Rungwe, southern Tanzania, which is not yet described.

Genus *Paracrocidura*

There are three species of *Paracrocidura* in Africa. *P. schoutedeni* occurs in Cameroon, Congo, Gabon, and Zaïre (Hutterer 1986). There is no indication that it is declining or that there is a need for a threatened species category. The other two species are of conservation concern:

Paracrocidura maxima - This species is known from a few specimens in the Albertine Rift area. Its exact range is unknown but it is probably restricted to eastern Zaïre, the Ruwenzori Mountains (in Uganda), and Rwanda.

Paracrocidura graueri - This species is known only from a single museum specimen which was collected in the Itombwe Mountains, eastern Zaïre.

Genus *Suncus*

Six species occur in Africa, of which *S. murinus* is introduced and *S. etruscus* also occurs in Europe and Asia. The remaining four species are confined to the African continent. Of these, only one is of immediate conservation concern:

Suncus remyi - This species is known from only two undisturbed forest areas in Gabon (Belinga and Makokou). Nothing is known of its ecology.

Genus *Sylvisorex*

There are eight species confined to Africa (including the island of Bioko in Equatorial Guinea). These shrews are usually considered to be forest species, though *S. megalura* also occurs in grasslands (Ansell 1960). There are three species of conservation concern:

Sylvisorex howelli - This species is endemic to the mountain forests of eastern Tanzania. It has been recorded from the Ulugurus and Usambaras.

Sylvisorex ollula - This species is known from just a few records in southern Cameroon (Bitye, Dja River) and nearby Gabon.

Sylvisorex vulcanorum - This recently described species (Hutterer and Verheyen 1985) is known from only seven specimens (Hutterer *et al.* 1987) from the Zaïre section of the Virunga Mountains and from Volcanoes National Park and Nyungwe Forest in Rwanda. Intensive collecting efforts in the Lake Kivu area have not yielded any further specimens and it almost certainly merits a threatened species category.

Genus *Ruwenzorisorex*

The genus *Ruwenzorisorex* includes just one species, *R. suncoides*. Although this species is sometimes assigned to *Sylvisorex*, it is in fact quite distinct. This species is apparently rare, known from only five specimens (Hutterer *et al.* 1987). It is endemic to the Albertine Rift, and is known from the Ruwenzori Mountains (Uganda), nearby eastern Zaïre, and Rwanda. It is therefore of great conservation concern.

Genus *Myosorex*

There are at least 13 species in the genus *Myosorex*, all confined to the African mainland and Bioko Island. The genus has a wide range in central-west, central, east, and southern Africa. One South African species, *Myosorex longicaudatus*, has been listed as Indeterminate by Smithers (1986) although Meester (pers. comm.) believes the species to be in no danger, as it has a relatively wide distribution. Five species of conservation concern, all with relatively restricted ranges, are listed below:

Myosorex schalleri - Known from a single specimen from the Itombwe Mountains, eastern Zaïre.

Myosorex polli - Known from a single specimen. The type locality is Kasai, Lubondai, Zaïre.

Myosorex eisentrauti - This species is confined to the western Cameroon mountains and Bioko Island, Equatorial Guinea.

Myosorex geata - There is only one record of this species, from the Uluguru Mountains, Tanzania.

Myosorex longicaudatus - First described in 1978 (Meester and Dippenaar 1978), this shrew is only known from forested areas in the southern parts of Cape Province in South Africa. It is little known but appears to favour fern clumps. It occurs in the Diepwalle Forest Reserve.

Genus *Scutisorex*

The single species, *Scutisorex somereni*, occurs in northern Zaïre, Uganda, and Rwanda (Meester and Setzer 1974; Kingdon 1974). There is no evidence to suggest the need for a threatened species category.

Chapter 4: The Elephant-Shrews

4.1 Introduction

The elephant-shrews comprise a relatively small number of generally well defined species that are confined to the continent of Africa. Although many of the species are active during at least part of the day, and are therefore relatively easy to see, they are not particularly well-known biologically. This lack of knowledge is, in part, due to their low population densities (despite wide geographical distributions) and cryptic habits (Brown 1964; Sauer and Sauer 1972; Rathbun 1979).

The rufous elephant-shrew (*Elephantulus rufescens*)
(Photo by G. B. Rathbun)

During the Miocene, the elephant-shrews were a much more diverse group than they are now, being represented by at least four subfamilies that included insectivorous as well as herbivorous forms (Patterson 1965). Presently, there are two subfamilies with 15 species (see Table 4.1). All species are principally insectivorous, although some are known to eat limited quantities of plant matter (Rathbun 1979). Most of the extant species are associated with specific habitats. For example, the eastern rock elephant-shrew (*Elephantulus myurus*) occurs almost exclusively on kopjes or large rock outcroppings (Critch 1969). The desert-dwelling short-eared elephant-shrew (*Macroscelides proboscideus*) is one of the smallest species (45g), while the forest-dwelling golden-rumped elephant-shrew (*Rhynchocyon chrysopygus*) is probably the largest, weighing approximately 550 g.

Most elephant-shrew species are not under threat of immediate extinction and none are listed in the 1988 IUCN Red List. A subspecies of the four-toed elephant-shrew (*Petrodromus tetradactylus beirae*) is considered Rare by Smithers (1986). However, this listing is based on its limited distribution in South Africa. Because this subspecies also occurs in Mozambique, it probably should not be considered threatened on a global basis. The status of most elephant-shrew species is undetermined. Several forms occupy very restricted forest habitats that are undergoing severe impacts from human development. The conservation of many of these "island forests" is critical owing to the special considerations of human encroachment (Brown 1981), island biogeographical theory (Diamond 1981), and population genetics (Lacy 1988). The forest-dwelling elephant-shrews (*Rhynchocyon* spp., and *Petrodromus* in some areas) may not be adaptable enough to survive the destruction of their habitat, and they will probably be locally extirpated with the disappearance of the fragmented and isolated forest patches they occupy. However, there is some evidence that *Rhynchocyon* species may adapt to altered habitats, provided there is suitable cover and plentiful invertebrates. In coastal Kenya, the golden-rumped elephant-shrew (*R. chrysopygus*) has been sighted in fallow cashew plantations (Rathbun, pers. obs.) and in the southern Uzungwa Mountains of Tanzania, the black-and-rufous elephant-shrew (*R. petersi*) forages in the gardens of tea plantation houses (J. Lovett, pers. comm.).

The forests that are important to the elephant-shrews are often characterized by numerous endemic plants (Hedberg and Hedberg 1968), as well as several populations of endemic or isolated mammals (Kingdon 1971; Kingdon 1981; Rodgers, et al. 1982) and birds (Moreau 1966; Collar and Stuart 1985). By developing and implementing conservation strategies for the elephant-shrews in these forests the overall biological diversity will be better protected in these unique ecosystems.

4.2 Classification

The most recent classification (Corbet 1971) is considered definitive, and is followed here (Table 4.1). Detailed taxonomic revisions of the elephant-shrews are found in Corbet and Neal (1965), Corbet and Hanks (1968) and Corbet (1971), including descriptions of subspecies and detailed distribution maps.

4.3 Biogeographic Review of the Elephant-shrews

The extant elephant-shrews are classified into four genera (Table 4.1) that are restricted to extreme north-western Africa and central, eastern, and southern Africa, south of the Sahara. The elephant-shrews are absent from West Africa and the Sahara region (Figure 4.1). The north African elephant-shrew (*Elephantulus rozeti*) is found only in north-western Africa; the dusky-footed elephant-shrew (*E. fuscipes*) is distributed over a relatively small area of eastern central Africa; the Somali elephant-shrew (*E. revoili*) is limited to Somalia; the rufous

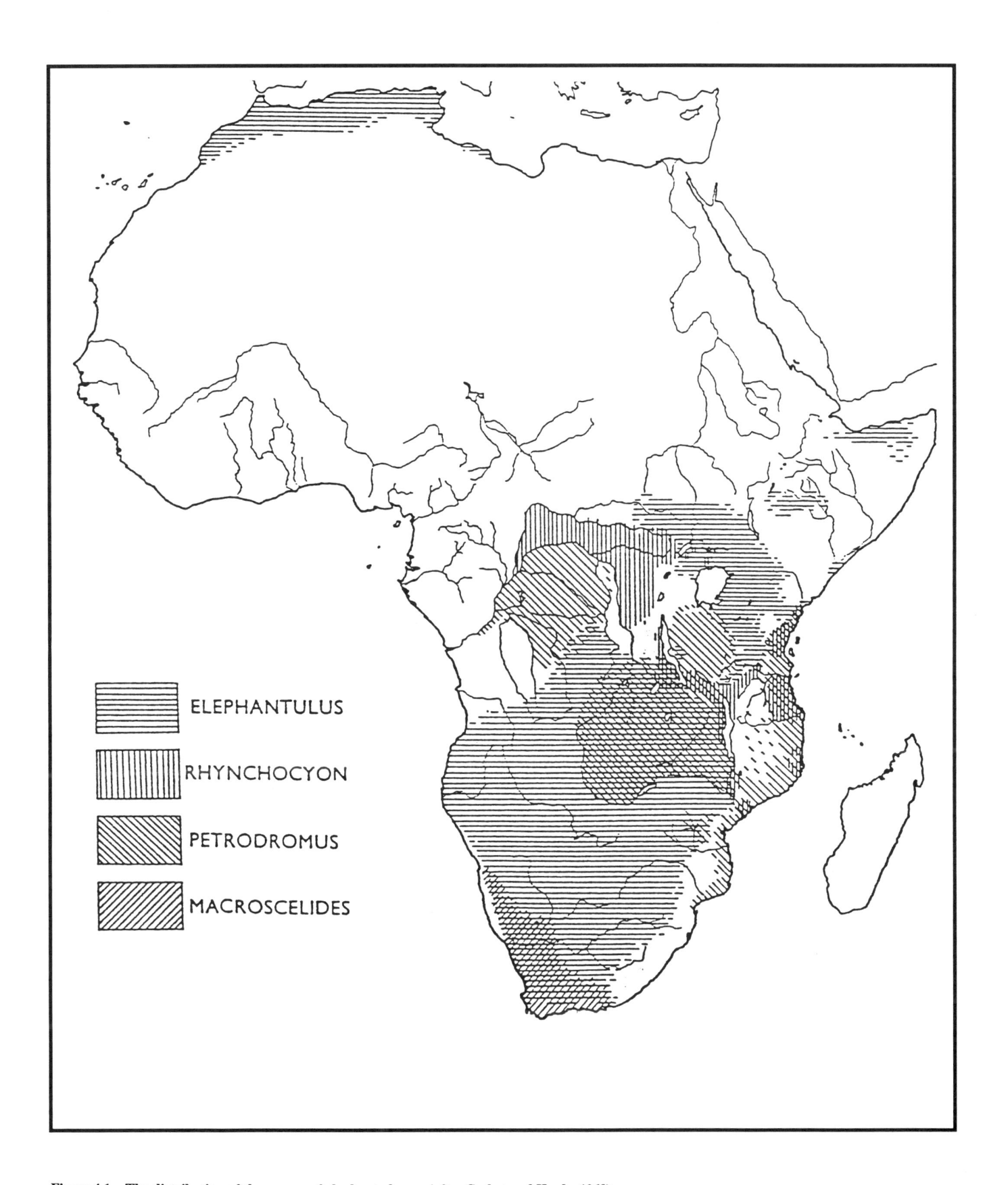

Figure 4.1 The distribution of the genera of elephant-shrews (after Corbet and Hanks 1968).

Table 4.1: Classification of the elephant-shrews. Number of subspecies in brackets, followed by distribution by country (based on Corbet and Hanks (1968) and Corbet (1971)).

Order: Macroscelidea

Family: Macroscelididae
Subfamily: Rhynchocyoninae
Genus: *Rhynchocyon*

R. chrysopygus Gunther, 1881 Golden-rumped elephant-shrew	[0]	Kenya
R. cirnei Peters, 1847 Chequered elephant-shrew	[6]	Central African Republic, Chad, Malawi, Mozambique, Tanzania, Uganda, Zaire, Zambia
R. petersi Bocage, 1880 Black-and-rufous elephant-shrew	[2]	Kenya, Tanzania

Subfamily: Macroscelidinae
Genus: *Petrodromus*

P. tetradactylus Peters, 1846 Four-toed elephant-shrew	[10]	Angola, Chad, Kenya, Malawi, Mozambique, Rwanda, South Africa, Tanzania, Zaire, Zambia, Zimbabwe

Genus: *Macroscelides*

M. proboscideus (Shaw, 1800) Short-eared elephant-shrew	[2]	Namibia, South Africa

Genus: *Elephantulus*

E. brachyrhynchus (A. Smith, 1836) Short-snouted elephant-shrew	[0]	Angola, Kenya, Mozambique, Namibia, Malawi, South Africa, Sudan, Tanzania, Uganda, Zaire, Zambia, Zimbabwe
E. edwardii (A. Smith, 1839) Cape elephant shrew	[0]	South Africa
E. fuscipes (Thomas, 1894) Dusky-footed elephant-shrew	[0]	Sudan, Uganda, Zaire
E. fuscus (Peters, 1852) Dusky elephant-shrew	[0]	Malawi, Mozambique, Zambia
E. intufi (A. Smith, 1836) Bushveld elephant-shrew	[0]	Angola, Botswana, Namibia, South Africa
E. myurus Thomas and Schwann, 1906 Eastern rock elephant-shrew	[0]	Botswana, Mozambique, Namibia, South Africa, Zimbabwe
E. revoili (Huet, 1881) Somali elephant-shrew	[0]	Somalia
E. rozeti (Ducernoy, 1830) North African elephant-shrew	[2]	Algeria, Libya, Morocco, Tunisia
E. rufescens (Peters, 1878) Rufous elephant-shrew	[0]	Ethiopia, Kenya, Somalia, Sudan, Tanzania, Uganda
E. rupestris (A. Smith, 1831) Western rock elephant-shrew	[0]	Botswana, Mozambique, Botswana, Namibia, South Africa, Zimbabwe

elephant-shrew (*E. rufescens*) is found in semi-arid areas of East Africa; the black-and-rufous elephant-shrew (*Rhynchocyon petersi*) and the golden-rumped elephant-shrew (*R. chrysopygus*) are both found in forested habitats, the golden-rumped elephant-shrew is restricted to coastal forests in Kenya, while the black-and-rufous elephant-shrew is also found in coastal forests (but not with overlapping ranges with the golden-rumped elephant-shrew) as well as on isolated montane forests further inland in both Kenya and Tanzania; the short-eared elephant-shrew (*Macroscelides proboscideus*), the rock elephant-shrew (*E. rupestris*), and the bushveld elephant-shrew (*E. intufi*) are confined to south-western Africa; the Cape elephant-shrew (*E. edwardi*) is found only in extreme southern Africa, whereas the eastern rock elephant-shrew (*E. myurus*) is distributed over a wider area of southern Africa; and the four-toed elephant-shrew (*Petrodromus tetradactylus*) probably has one of the widest distributions, from central and eastern Africa south to the south-eastern part of the continent. The chequered elephant-shrew (*R. cirnei*) has a similar distribution to the four-toed elephant-shrew, but not as westerly or southerly. The short-snouted elephant-shrew (*E. brachyrhynchus*) has a similarly wide distribution in central Africa, whereas the dusky elephant-shrew (*E. fuscus*) has a more restricted distribution in central Africa. In summary, the greatest diversity of elephant-shrews is found in south-western and eastern Africa (see Tables 1.1 and 4.1).

The genera *Macroscelides* and *Elephantulus* are restricted to lowland semi-arid savannahs, bushlands, and woodlands, while *Rhynchocyon* spp. are found in lowland and montane forests and dense woodlands. *Petrodromus* occupies dense scrub and woodlands as well as forests, often overlapping ranges with *Elephantulus* or *Rhynchocyon*. The forest-dwelling forms of *Rhynchocyon* have a highly disjunct distribution that closely matches the patchy distribution of forests in eastern Africa (Figures 4.2 and 4.3).

The six vulnerable or threatened forest-dwelling elephant-shrews (representing four full species: Table 4.2) occur in five distinct forest ecosystems:

1. "Coastal East Africa" including the lowland, semi-deciduous dry forests along the coast of Kenya, Tanzania and Mozambique from the Boni Forest to the Pugu Forest and south into Mozambique (Figure 4.3; Moomaw 1969; Stuart 1981).

2. The "Zanzibar forests" that occur on the islands of Zanzibar and Mafia (Figure 4.3; Silkiluwasha 1981) are wetter and more evergreen than those that occur along the coast of the mainland.

These two lowland coastal forest habitats support four of the six threatened forms: the golden-rumped elephant-shrew (*R. chrysopygus*), a subspecies of the chequered elephant-shrew *R.c. cirnei*, and both subspecies of the black-and-rufous elephant-shrew *R.p. petersi*, and *R.p. adersi*. Because little is known about *R.c. cirnei*, it is included here, although its preferred habitat may be woodland rather than true forest.

Table 4.2: Status of species and subspecies of elephant-shrews.
Ratings for each of the three categories are between 1 and 3, with 3 being most critical. ? = indeterminate, which carries a numerical value of one. The total is additive. Score of 9 = Endangered; 5-8 = Intermediate levels of threat, and 3 & 4 = safe, as far as is known.

Taxon	Status of Taxon	Uniqueness of Taxon	Status of Habitat	Total
Genus: *Rhynchocyon*				
R. chrysopygus	2	3	2	7
R. cirnei cirnei	?	1	3	5
R. c. hendersoni	?	1	3	5
R. c. macrurus	?	1	2	4
R. c. reichardi	?	1	1	3
R. c. shirensis	?	1	2	4
R. c. stuhlmanni	?	1	1	3
R. petersi adersi	?	2	2	5
R. p. petersi	?	2	3	6
Genus: *Petrodromus*				
P. tetradactylus beirae	?	1	1	3
P. t. rovumae	?	1	1	3
P. t. sangi	?	1	3	5
P. t. schwanni	?	1	2	4
P. t. sultan	?	1	2	4
P. t. swynnertoni	?	1	2	4
P. t. tetradactylus	?	1	1	3
P. t. tordayi	?	2	1	4
P. t. warreni	?	1	1	3
P. t. zanzibaricus	?	1	2	4
Genus: *Macroscelides*				
M. proboscideus flavicaudatus	?	2	1	4
M. p. proboscideus	?	2	1	4
Genus: *Elephantulus*				
E. brachyrhynchus	?	1	1	3
E. edwardi	?	1	1	3
E. fuscipes	?	1	1	3
E. fuscus	?	1	1	3
E. intufi	?	1	1	3
E. myurus	?	1	1	3
E. revoili	?	1	1	3
E. rozeti deserti	?	2	1	4
E. r. rozeti	?	2	1	4
E. rufescens	?	1	1	3
E. rupestris	?	1	1	3

3. "Eastern Arc Forests" occur on isolated, ancient crystalline mountains, up to 200 km inland from the East African coast. They extend from the Taita Hills in Kenya to the Uzungwa Mountains of Tanzania (Figure 4.3; Lovett 1985; Lovett and Thomas 1988; Lovett, *et al.* 1988; Rodgers and Homewood 1982a; Rodgers and Homewood 1982b). Although Mount Meru, Tanzania, and the Chulu Hills, Kenya, are volcanic in origin and are not technically part of the Eastern Arc Forests, they are included here for convenience. Several of the highly disjunct Eastern Arc Forests support subspecies of either the four-toed elephant-shrew (*P.t. sangi*), black-and-rufous elephant-shrew (*R.p. petersi*), or chequered elephant-shrew (*R.c. reichardi*).

4. The "Rift Valley System", which includes evergreen forests on isolated mountains and plateaus from the northern end of Lake Tanganyika south to the southern end of Lake Malawi (Figure 4.3), supports one vulnerable subspecies of the chequered elephant-shrew: *R.c. hendersoni*. Several other *Rhynchocyon* and *Petrodromus* subspecies occur in lowland and riverside forests and woodlands associated with the Rift Valley ecosystem, but presently they are not considered threatened. However, a survey of their habitats might change this assessment since these forests are known to be suffering from encroachment.

5. The final ecosystem includes isolated patches of "Central African Lowland Forest" in Uganda (Figure 4.3; Struhsaker 1981). Even though the chequered elephant-shrew subspecies *R.c. stuhlmanni* is widespread in Zaïre, and thus probably not vulnerable or threatened as a taxon, the isolated populations in Uganda may be vulnerable owing to habitat loss.

4.4 Taxa of Conservation Concern

Each species and subspecies has been rated on a scale of 1 to 3 (3 being most critical) for each of three criteria: a) threat of extinction; b) distinctiveness of the taxon within the elephant-shrews; and c) distinctiveness, abundance, and status of habitat. An overall score for each elephant-shrew taxon was then calculated by summing the ratings for the three categories. A score of 9 indicates a critical danger of extinction (IUCN

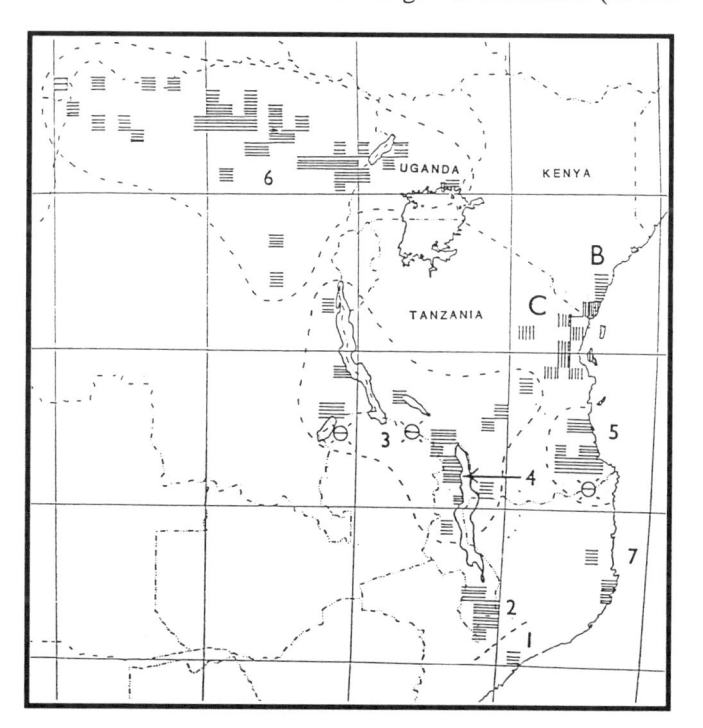

Figure 4.2 The distribution of *Rhynchocyon* species and subspecies in eastern Africa (from Corbet and Hanks 1968). B = *R. chrysopygus*, C = *R. petersi*, 1 = *R. c. cirnei*, 2 = *R. c. shirensis*, 3 = *R. c. reichardi*, 4 = *R. c. hendersoni*, 5 = *R. macrurus*, 6 = *R. c. stuhlmanni*, 7 = *R. c. subsp.* Circle = locality not precisely known. Square = record unconfirmed.

The golden-rumped elephant-shrew (*Rhynchocyon chrysopygus*)
(Photo by G. B. Rathbun)

category = Endangered), scores between 5 and 8 show intermediate levels of threat (IUCN categories Vulnerable, Rare, Indeterminate and Insufficiently Known), and 3 and 4 indicate no danger at present. Because little is known about many of the elephant-shrews and their habitats, a value of 1 was arbitrarily assigned when insufficient information was available for any particular criterion (see Table 4.2).

Of the 32 elephant-shrew taxa considered valid by Corbet and Hanks (1968) and Corbet (1971), six are under varying degrees of threat (IUCN categories Vulnerable, Rare, Indeterminate, or Insufficiently Known; Table 4.2). None are presently considered Endangered according to the IUCN categories of threat. All of the forms that appear to be under threat occur in fragmented or isolated forest patches where there is concern about the status of the habitat. The numerical score of some of these taxa might increase after surveys have been carried out, because the "insufficient information" status (with a minimal rating of 1) might be increased.

All of the species and subspecies of *Macroscelides* and *Elephantulus* have distributions or habitat requirements that indicate neither restricted numbers nor habitat vulnerability (Smithers 1971, 1983, 1986; Aulagnier and Thevenot 1986; Meester *et al.* 1986; Ansell and Dowsett 1988).

Golden-rumped Elephant-shrew (*Rhynchocyon chrysopygus*)

IUCN Category of Threat: Vulnerable.

This species is endemic to Kenya and occurs in forest patches north of Mombasa as far as the Boni Forest (Corbet and Hanks 1968; Rathbun 1979). The Arabuko-Sokoke Forest is a particularly important site for the species. Many of these coastal forests are under pressure from forestry practices (tree and pole cutting) and encroachment for agriculture. The species is trapped and eaten by local people, which might be affecting their numbers. These activities, together with its restricted distribution, are the basis of the serious concern for this species. A conservation proposal on this species is presented in Appendix 3.

Chequered Elephant-shrew (*Rhynchocyon cirnei*)

The chequered elephant-shrew is not threatened as a species, but at least two of its six subspecies appear to be so.

R.c. cirnei is only known from the type (and only) specimen from Quelimane in Mozambique, north of the Zambezi River

(Corbet and Hanks 1968), and the status of the species and its habitat is uncertain. IUCN Category of Threat: Insufficiently Known.

R.c. hendersoni has only been found in Malawi (three specimens) in a small, isolated montane forest along the west side of Lake Malawi (Corbet and Hanks 1968; Kingdon 1974; W.F.H. Ansell, pers. comm.). The status of the chequered elephant-shrew in this area is not known, but in view of its very restricted distribution, a threatened species category is warranted. IUCN Category of Threat: Rare.

R.c. stuhlmanni is widespread in Zaïre, and thus probably not vulnerable or threatened as a taxon, but there are isolated populations in Uganda which may be vulnerable due to habitat fragmentation. These isolated patches of lowland West African forest include Bwamba (Semliki), Bugoma, Budongo, and Mabira (Corbet and Hanks 1968; Kingdon 1974). In addition, it is not known whether *R.c. stuhlmanni* occurs in the forests south of Bwamba (Itwara, southern Kibale, and the Kasyoha-Kitomi/Kalinzu/Maramagambo complex), since the altitude of these forests might be too high. IUCN Category of Threat: not threatened.

R.c. reichardi and *R.c. shirensis* are found in isolated forests occurring on mountains and plateaus near Lakes Tanganyika and Malawi. There is little information on their current status (Ansell and Ansell 1973; Ansell and Dowsett 1988). In Tanzania, likely sites associated with Lake Malawi include forests near Rungwe, Poroto, Kipengere, and Matengo. Likely forests that are associated with Lake Tanganyika include the Gombe, Kungwe, Karema, and Mbizi forests. In Zambia and Malawi *R.c. shirensis* may occur in forests and woodlands associated with the Mafinga Mountains and Nyika Plateau (which is a national park) and also the Makutu Mountains in Zambia. IUCN Category of Threat: not threatened.

Black-and-rufous Elephant-shrew (*Rhynchocyon petersi*)

IUCN Category of Threat: Rare.

There are two subspecies, both of which are threatened. *R.p.petersi* is known from the coastal forests of Kenya (south of Mombasa) and Tanzania and from isolated montane forests further inland in both Kenya and Tanzania. *R.p. adersi* occurs on Zanzibar and Mafia Islands. Apparently *R.p. adersi* occurs in the Muungwi forest, Muyuni coastal strip forest, on Uzi Island, and possibly in the Jozani Forest on Zanzibar (Silkiluwasha 1981). It is unlikely that elephant-shrews occur in the groundwater forest at Jozani if it is prone to seasonal flooding (Rathbun, pers. obs).

Both subspecies occur in forest habitats which are threatened by human development. Deforestation is occurring owing to tree and pole cutting and also due to conversion to agricultural land. It may be that these elephant-shrews can adapt to modified habitats, but this has not been determined. The extent of the threat due to habitat change is therefore unclear. Elephant-shrews are eaten by local people along the

A juvenile four-toed elephant-shrew (*Petrodromus tetradactylus*)
(Photo by G. B. Rathbun)

Kenya coast (Rathbun, pers.obs.), which might have an impact on their numbers, although no quantitative data exist on this subject. A conservation proposal on this species is presented in Appendix 3.

Four-toed Elephant-shrew (*Petrodromus tetradactylus*)

This species has a wide distribution and is not threatened. However, one of its ten subspecies is threatened.

P.t. sangi has been reported only from the Taita Hills, Kenya. The small, isolated forest habitat in the Taita Hills is coming under increasing pressure from agriculture (Collins and Clifton 1984). However, there is a remote possibility that this subspecies has a wider distribution. The location on a specimen in the British Museum (Natural History) is "Kibwezi" (Corbet and Hanks 1968). It is not known if this site refers to the woodland and forest habitats below the Chyulu Hills near Kibwezi, Kenya, or the site with the same name on Mount Meru, Tanzania. Because of the gravity of the situation on the Taita Hills, and the unknown status elsewhere, there is considerable concern for this subspecies. IUCN Category of Threat: Insufficiently Known.

4.5 Captive Breeding of Elephant-shrews

Of the 15 elephant-shrew species, eight have been bred at least once in captivity. Only four of these species, however, have been kept and bred over any sustained length of time (Table 4.3). Significantly missing from the list of successfully bred genera is *Rhynchocyon* (which includes most of the threatened taxa), and captive accomplishments with the four-toed elephant-shrew have been limited. Until recently, there was little success in keeping any of the elephant-shrews in captivity, owing to their poorly understood biology (especially their social behaviour and mating systems); small numbers in founder

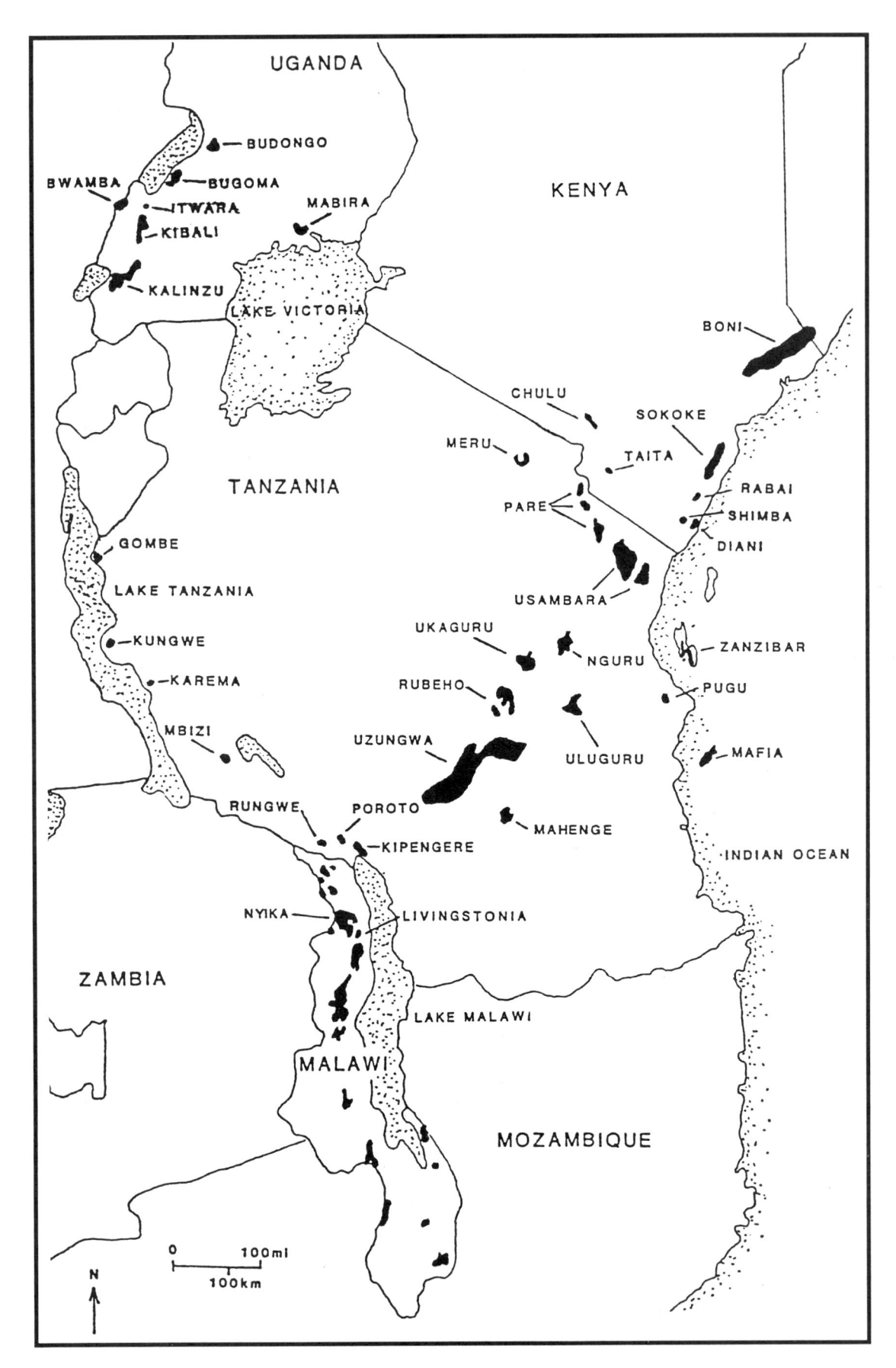

Figure 4.3 The major forest blocks of eastern Africa that are important habitat for selected elephant-shrews (see text). Adapted principally from Ansell and Dowsett (1988), Howell (1981), Lovett (1985), Struhsaker (1981), and Stuart (1981).

Table 4.3: Reports concerning attempts to maintain elephant-shrews in captivity.

Taxon	Citation
Rhynchocyon chrysopygus	Rathbun et al. 1981; Frankfurt Zoo, unpubl. data
Petrodromus tetradactylus*	Crandall 1964; Ansell and Ansell 1969; Tripp 1972
Macroscelides proboscideus*	Pocock 1912; Rosenthal 1975
Elephantulus brachyrhynchus*	Rankin 1965, 1967
E. intufi*	Hoesch 1959; Tripp 1972
E. myurus*	Horst 1946, 1954; Critch 1969; Tripp 1972
E. rozeti*	Flower 1931; Dachsel 1964; Tripp 1972; Olney 1979; Seguignes 1983; Rosenthal, pers. comm.
E. rufescens*	Hoogstraal 1950; Walker 1955; Hoopes and Montali 1980; Houck, *et al.* 1981; Rathbun et al. 1981; Lumpkin et al. 1982; Keppel 1985; Koontz 1984; Lumpkin 1986; Lumpkin and Koontz 1986
E. rupestris	Pocock 1912; Olney 1979; Rosenthal pers. comm.

Notes: Taxa for which there are no data are not listed.
*Species which have reproduced in captivity.

stocks; inadequate enclosure designs; and poor diets for captive animals. Fortunately, advances in zoo biology and new information on their natural history have resulted in the founding of several successful elephant-shrew colonies (Rosenthal 1975; Houck et al. 1981; Rathbun et al. 1981; Keppel 1985; Lumpkin 1986), including a group of rufous elephant-shrews (*E. rufescens*) at the National Zoological Park (Washington, D.C.) that has bred successfully for seven generations (F. Koontz, pers. comm.).

Judging from the recent successes in maintaining and breeding *Elephantulus* and *Macroscelides* (Table 4.3), the husbandry techniques for these forms are reasonably well formulated. Improving husbandry procedures for *Petrodromus* and *Rhynchocyon*, however, is necessary if captive propagation is ever to be used to assist in their conservation. Because the four-toed elephant-shrew subspecies *P.t. sangi* may be found to be threatened or even endangered when field work is completed (see section 5.3.9 Kenya), it may be necessary to build upon current husbandry techniques for this taxon, if captive breeding is to be accomplished. *Rhynchocyon* has been difficult to maintain in captivity and has never bred under such circumstances (Kingdon 1974; Rathbun 1979). However, recent improvements in animal care, and the single golden-rumped elephant-shrew (*R. chrysopygus*) that has been kept successfully at the Frankfurt Zoo, suggest that a renewed effort with this species should be considered (F. Koontz and M. Rosenthal, pers. comm.).

If captive propagation is ever going to be used to increase numbers of threatened elephant-shrews for re-introductions, then research in husbandry and breeding methods for *Petrodromus* and *Rhynchocyon* will have to be initiated before populations become critically low. Unfortunately, few zoos devote space to elephant-shrew propagation, and even fewer have conducted any noteworthy research on these species (important exceptions being the National Zoological Park and the Lincoln Park Zoo, Chicago). However, *Rhynchocyon* and *Petrodromus* are striking and unusual mammals and they have a high zoo exhibit value. If the husbandry methods could be developed at one or two zoos, then the resulting number of institutions that might be willing to contribute holding space to breeding programs could be significant (F. Koontz, pers. comm.). Captive propagation, however, should be considered as only one component of a conservation strategy aimed primarily at protecting the species' habitat.

Chapter 5: Conservation Action Plan

5.1 Rationale for Insectivora and Elephant-shrew Conservation

From the preceding analysis, it is clear that the conservation of the African Insectivora and elephant-shrews will only succeed if it is fully integrated into the broader issues of development and environmental management on the continent and its surrounding islands. These species do not have a particularly strong public appeal (though we hope that this Action Plan will help to promote greater interest in these remarkable animals), and so it is difficult to raise support for their conservation. However, this need not be a serious problem since it appears that the threats they face are generally not species-specific. There is little evidence to suggest that any of these species is at risk from direct exploitation. Rather, they are threatened by habitat clearance which affects most species, and changing land-use patterns resulting in soil loss (thereby threatening golden moles) or in the pollution and siltation of rivers (threatening the otter-shrews and the aquatic tenrec).

This Action Plan emphasises two immediate types of action that are important for the African Insectivora and elephant-shrews: habitat conservation; and an immediate need for additional field surveys. The first is essential if the full diversity of these species is to survive. The second is important since, without more knowledge, we cannot be certain that the efforts to conserve the most diverse habitats are addressing the highest priorities for these species. The need for habitat conservation in Africa and Madagascar is so urgent that it cannot wait for surveys to be carried out. However, the surveys should also start as soon as possible, since otherwise important areas might be lost before it is known whether or not they are of significance for Insectivora and elephant-shrew conservation. In Appendices 1-3, three brief proposals that are of particularly high priority for the Insectivore, Tree Shrew and Elephant Shrew Specialist Group are presented (on the aquatic tenrec, golden moles and elephant-shrews).

The Action Plan also identifies two subsidiary forms of conservation action that are urgently needed: specific research projects; and captive breeding programmes. Although possibly less immediately pressing than either habitat or field surveys, they are nevertheless essential for certain species.

The conservation of Africa's Insectivora and elephant-shrews will have much wider benefits for the conservation of biological diversity generally. Distributional analyses clearly show that the highest priority areas for these species are the Pleistocene refugia that have already been identified as being important for a number of species of endemic primates, rodents, birds, amphibians, invertebrates and plants. The recommendations made in this chapter are therefore supportive of global priorities for the conservation of biological diversity.

5.2 Habitat Protection

In general, it seems that widely distributed savanna species may adapt readily to cultivated lands, forestry plantations, and settlements (Hutterer et al. 1987), and are least likely to require specific conservation projects. Forest species, however, tend to be dependent on the existence of natural forests. This is particularly critical for forest species with limited ranges whose habitats are being fragmented through human encroachment. Examples include golden moles in southern Africa (Meester 1976; Smithers 1986), the 'central African endemic' shrews (Hutterer et al. 1987), and some of the elephant-shrews in eastern Africa. Species adapted to arid and semi-arid conditions vary in the level of their conservation requirements. For instance, some widely distributed species, such as the Algerian hedgehog (*Atelerix algirus*), are capable of adapting to human-induced impacts on land-use, in contrast to some golden moles with highly specific habitat requirements, limited ranges, and which require undisturbed desert conditions.

It is important to determine to what extent threatened or potentially threatened species of Insectivora and elephant-shrews are present in existing protected areas. The faunas of many protected areas have already been described as a result of surveys, often as a prelude to management planning. However, many reserves, particularly in moist forest environments or those that have recently been created, have not been surveyed, or have only been surveyed for particular species such as large mammals or birds. It is strongly recommended that small mammal work be included in all surveys and inventories carried out in protected areas, using techniques appropriate for Insectivora and elephant-shrews. Where lethal trapping is not permitted, live trapping may be carried out as an alternative. However, since this might not be effective in capturing all species present, and since identification often requires museum specimens for comparisons, parallel specimen collection in adjacent similar habitats is recommended where deemed necessary.

There are a number of species with restricted ranges that are unlikely to occur in any existing effective protected areas. Examples may include the aquatic tenrec (*Limnogale mergulus*), some of the shrew-tenrecs *Microgale* spp., several of the central African endemic shrews (Hutterer et al. 1987), both of the lesser otter-shrews *Micropotamogale* spp., a number of South African golden moles (Meester 1976; Smithers 1986), and some of the East African elephant-shrews. In many cases, the natural environments where these animals are known to occur are clearly under threat from encroachment, forestry exploitation, or mining. Since these are also critical areas for biological diversity generally, it is a high priority to inform the relevant authorities and decision-makers of their international responsibility to protect or manage rationally sizeable portions of the remaining natural habitats.

Before proceeding to site-specific recommendations, a couple of cautionary notes are in order. Firstly, even if the habitat of a species is adequately conserved, this does not necessarily mean that the species is safe. Species can be extirpated locally as a result of environmental fluctuations, especially if their populations are small and fragmented. Small populations are much more vulnerable to the unexpected or uncontrollable, such as an outbreak of disease, or habitat removal through fire, or illegal habitat clearance in a reserve during a period of breakdown in law and order. Also, small populations are much more vulnerable to the deleterious effects of inbreeding for both genetic and demographic reasons. These effects can show themselves as decreased breeding success and increased vulnerability to disease. It is therefore important that habitat conservation programmes aim to conserve suitable areas of sufficient size to contain populations of Insectivora and elephant-shrews that are large enough to be viable in the long-term. There is no magical number for this "minimum population size" (it varies between species), but most geneticists advise that alarm bells should start ringing for populations of less than 500 animals. If possible, conservationists should aim to spread these risks and attempt to conserve several populations greater than 1,000 animals of each threatened species. Obviously, in the case of the African Insectivora and elephant-shrews, we do not yet know enough about their population densities to know whether or not we are meeting these targets, but this indicates a valuable area for future research (see section 5.4).

Secondly, it has become increasingly clear in recent years that field conservation programmes can only be successful in the long-term if local people are in support of them. The driving force behind much of the ecological degradation in Africa and Madagascar is poverty: this poverty itself drives people to use their land in ways that produce short-term benefits, even though this may be very damaging in the long-term. Habitat conservation efforts should therefore be coupled with efforts to improve the living conditions for local people. There are many ways in which this can be done, but all should have the aim of linking biological conservation with rural development. In certain instances, a strict nature reserve or national park serves only to antagonize local people. In many areas, the long-term future of habitats and species can be made more secure through permitting the sustainable use of certain resources within reserves, thereby allowing people to obtain direct benefits from conservation. Strict reserves in which all forms of utilisation are prohibited, though certainly necessary in some areas, are not always the most effective means of obtaining conservation objectives, and sometimes more integrated approaches to conservation can be more effective.

The following specific habitat conservation activities are recommended for the African Insectivora and elephant-shrews. This section is organised on a country by country basis.

5.2.1 Cameroon

Conservation of forests in the western highlands
Three shrew species of conservation concern occur in the western highlands: *Crocidura eisentrauti* is endemic to Mount Cameroon; *C. manengubae* to Mount Manenguba; and *Myosorex eisentrauti*, which occurs more widely in the highlands, as well as on Bioko, Equatorial Guinea. More effective forest conservation measures are needed for Mount Cameroon, and especially on Mount Manenguba, where the remaining forest is severely threatened. In this last site, conservation measures will require integrated rural development programmes, such as those already being carried to the north on Mount Oku (or Kilum) by the International Council for Bird Preservation (ICBP).

Conservation of the Dja Game Reserve
This reserve is one of the most important sites for biological diversity in southern Cameroon, and its conservation will secure the future of the shrew *Sylvisorex ollula*, otherwise known only from northern Gabon.

5.2.2 Central African Republic

Conservation of forest habitats in the Boukoto region
The shrew *Crocidura grassei* is only known from this site and one locality in Gabon, and may be at risk.

5.2.3 Equatorial Guinea

Conservation of high altitude forests on Bioko
The rare shrew *Myosorex eisentrauti* occurs in this site, and otherwise only in the Cameroon highlands.

5.2.4 Ethiopia

Conservation of forests east of the Rift Valley
Forest conservation is needed for six rare shrews: *Crocidura glassi* (Gara Mulata Mountains); *C. lucina* (high altitudes around Dinshu); *C. phaeura* (Mount Gwamba in Sidamo Province); *C. thalia* (north-eastern Bale Province); and two undescribed *Crocidura* species in the Harrenna Forest of the Bale Mountains National Park. The continued effective protection of this last mentioned site is of particular importance. For those threatened species that do not occur there, other protected areas or integrated rural development programmes will probably be needed.

Conservation of forests and moorland west of the Rift Valley
The shrew *Crocidura baileyi* is endemic to these moorlands. Its habitat might be conserved in Simien National Park where security problems are preventing active management at the moment.

5.2.5 Gabon

Conservation of forests in northern Gabon
Four rare species of shrew occur in this area, especially around Makkokou and Belinga: *Crocidura crenata* (also in Zaire); *C. grassei* (also in Central African Republic); *Suncus remyi* (apparently endemic); and *Sylvisorex ollula* (also in southern Cameroon).

5.2.6 Guinea

Conservation of Mount Nimba
Two threatened species occur, the Nimba otter-shrew (*Micropotamogale lamottei*), and the shrew *Crocidura nimbae*. Both are endemic to the mountain (also occurring in the Liberian and Ivorian sectors). The proposed Biosphere Reserve should be established, and mining operations should be prevented.

5.2.7 Ivory Coast

Conservation of Mount Nimba.
Two threatened species occur, the Nimba otter-shrew (*Micropotamogale lamottei*), and the shrew *Crocidura nimbae*. Both are endemic to the mountain (also occurring in the Liberian and Guinean sectors). Forest conservation measures should be reinforced, and mining operations should be prevented.

Conservation of moist savanna habitat
The shrew *Crocidura wimmeri* is endemic to moist savanna around Adiopodoume in Ivory Coast and suitable sites need to be conserved.

5.2.8 Kenya

Conservation of coastal forests
The golden-rumped elephant-shrew (*Rhynchocyon chrysopygus*) is endemic to the numerous widely-spaced forests along the coast of Kenya from Mombasa to near the Somali border. South of Mombasa, the black-and-rufous elephant-shrew (*R. petersi*) occurs in isolated forests. Only some of the coastal forests where these elephant-shrews occur have any protection, and often these areas are very small. The protected areas include a part of the Shimba Hills Reserve, a small part of the Arabuko-Sokoke Forest Reserve, Gedi Historical Monument, and a small portion of the Boni Forest. Other "island" forests enjoy *de facto* protection because they are sacred "kayas" of the coastal people. All these sites, in particular the larger forest areas, need to be conserved for their elephant-shrews. Management practices should be reviewed with the goal of increasing protected elephant-shrew habitat. These activities should be coordinated with the National Museums of Kenya Coastal Forest Survey. See Appendix 3 for more information.

Conservation of the Taita Hills
The small patches of forest on the Taita Hills should continue to be protected for the rare subspecies of the four-toed elephant-shrew (*Petrodromus tetradactylus sangi*) (which is not known for certain from anywhere else).

Conservation of forests in western Kenya
The rare shrew *Crocidura monax* occurs in these forests (otherwise only in Tanzania). Appropriate conservation measures need to be assessed.

Conservation of forest on Mount Gargues
Mount Gargues is the only known locality of the shrew *Crocidura raineyi*, which is dependent on the continued protection of its forest habitat.

5.2.9 Liberia

Conservation of Mount Nimba
Two threatened species occur, the Nimba otter-shrew (*Micropotamogale lamottei*), and the shrew *Crocidura nimbae*. Both are endemic to the mountain (also occurring in the Ivorian and Guinean sectors). The small amount of forest remaining in Liberia should be conserved and no further expansion of mining operations should be permitted.

5.2.10 Madagascar

For Madagascar, there are a very large number of sites in need of protection:

Conservation of Sambirano forests
The Sambirano forests form a small distinctive mainland forest isolate in the north-west, extending to the island of Nosy Be. The largest undisturbed block lies on the plateau of Manongarivo which is a Special Reserve and probably the most important site for habitat conservation.

Conservation of Marojejy Massif
The Marojejy Massif is a Strict Nature Reserve (Réserve Naturelle Intégrale 12) situated between Andapa and Sambava in the north-east, and requires continued protection.

The golden-rumped elephant-shrew (*Rhynchocyon chrysopygus*)
(Photo by G. B. Rathbun)

A shrew-tenrec species (*Microgale talazaci*)
(Photo by M. E. Nicoll/BIOS)

Conservation of Masoala Peninsula
The Masoala Peninsula lies in the north-east and has a high level of local plant and animal endemism. It is currently proposed as a national park area, and this should be put into effect. The newly discovered dark pigmy shrew-tenrec (*Microgale pulla*), was recently described from the opposite side of the Antongil Bay and might occur on the peninsula.

Conservation of Ambatovaky Special Reserve
This reserve comprises the largest protected extent of Eastern Domain lowland rainforest. Two relatively little-known tenrecs, the striped shrew-tenrec (*Microgale melannorachis*) and the lesser long-tailed shrew-tenrec (*M. longicaudata*), both occur.

Conservation of Ranomafana Est region
A large forest block near Ranomafana Est is currently being proposed as a national park, and this should be implemented. A range of rare bird and mammal species have been recently recorded there, including the aquatic tenrec (*Limnogale mergulus*) (see Appendix 1).

Conservation of Upland rainforest isolates
There are several isolated rainforest blocks on the central plateau and to the north of the main eastern rainforest block. Important sites for protection are the Ambohitantely Special Reserve 130 km north of Antananarivo, the Ambohijanahary Special Reserve 300 km west of Antananarivo on the western edge of the central plateau, and in Montagne d'Ambre National Park in the extreme north. Two *Microgale* species, northern lesser long-tailed shrew-tenrec (*M. prolixacaudata*) and Drouhard's shrew-tenrec (*M. drouhardi*), both occur in the extreme north.

Conservation of Tsimanampetsotsa
This strict nature reserve (R.N.I. 10) lies in the south-west, and supports spiny thicket dominated by the endemic Didieriaceae on limestone karst, and sand dune/alkaline lake fringe vegetation. It is one of the few shrew-tenrec localities outside the Eastern Centre of Endemism.

Conservation of Zombitse and Vohibasia Forests
These two forests remain as isolated pockets of western deciduous forest in the south-west. They are unusual within this region in that the greater hedgehog-tenrec (*Setifer setosus*) occurs there, while normally absent in the south-west. The large-eared tenrec (*Geogale aurita*) is also present.

Conservation of Forest concession Centre de la Formation Professionelle Forestière, Morondava
This natural forest north of Morondava is currently exposed to low-level logging and has excellent facilities and access. It is an important site for conservation, since shrew-tenrecs referrable to either the short-tailed shrew-tenrec (*M. brevicaudata*) or the western short-tailed shrew-tenrec (*M. occidentalis*) occur, and possibly also the lesser long-tailed shrew-tenrec (*M. longicaudata*).

Conservation of Bemaraha and the Bemamba and Masama Lakes Region
The calcareous Bemaraha Strict Nature Reserve (Réserve Naturelle Intégrale 9) supports a wide range of locally endemic vertebrate species. The forests and wetlands around Lakes Bememba and Masama may also prove to be important sites for insectivores endemic to the western part of the island.

Conservation of Ankarafantsika
The Ankarafantsika massif with its extensive deciduous forests and small lakes might hold little-known furred tenrec species.

5.2.11 Mozambique

Conservation of the habitat of the yellow golden mole
The yellow golden mole (*Chlorotalpa obtusirostris*) occurs in alluvial soils, dry river beds and sand dunes in the south of the country, as well as in neighbouring South Africa and Zimbabwe; its habitat is probably in need of protection.

5.2.12 Namibia

Conservation of coastal sand dune habitat in south of country
Grant's golden mole (*Eremitalpa granti*) occurs in sand dunes in southern Namibia, as well as in neighbouring South Africa. Its conservation depends in part on the continued protection of the Namib Desert National Park. See Appendix 2 for more details.

5.2.13 Nigeria

Conservation of swamps near Bahindi
The shrew *Crocidura longipes* is known only from two swamps east of Bahindi near Kainji Lake National Park. These sites should be conserved, perhaps as part of an integrated rural development programme.

5.2.14 Rwanda

Conservation of the Nyungwe Forest Reserve and Volcanoes National Park
Four rare species of shrew occur in forests in Rwanda, with the above two sites being the most important. The species are: *Crocidura lanosa, Paracrocidura maxima, Sylvisorex vulcanorum,* and *Ruwenzorisorex suncoides*. All these species also occur in neighbouring Zaire, and two of them in Uganda.

5.2.15 Sao Tomé and Príncipe

Conservation of forests on Sao Tome island
The very rare endemic shrew *Crocidura thomensis* depends of the forests of Sao Tomé for its continued survival.

5.2.16 South Africa

Conservation of the Diepwalle Forest Reserve
This is a key locality for the shrew *Myosorex longicaudatus* and should continue to receive protection, along with other localities for the species in southern Cape Province.

Establishment of Groen River National Park
Grant's golden mole (*Eremitalpa granti*) occurs in no protected area in South Africa, but does in the proposed Groen River National Park, which should be established. The species also occurs in Namibia. See Appendix 2 for more details.

Conservation of golden mole habitat in the coastal areas of eastern Cape Province, Transkei and Natal
Three threatened species of golden mole occur in this area, and all require habitat conservation action: Duthie's golden mole (*Chlorotalpa duthiae*) (restricted to Eastern Cape); giant golden mole (*Chrysospalax trevelyani*) (Eastern Cape and Ciskei); and Zulu golden mole (*Amblysomus iris*) (occurring more widely in this coastal region). The giant golden mole also needs more parts of its habitat fenced to protect it from dogs. See Appendix 2 for more details.

Conservation of the habitat of the yellow golden mole
The yellow golden mole (*Chlorotalpa obtusirostris*) occurs in alluvial soils, dry river beds and sand dunes in north-east Transvaal, as well as in neighbouring Zimbabwe and Mozambique; its habitat is probably in need of protection.

Conservation of sand dunes at Port Nolloth
These dunes are the only known site for De Winton's golden mole (*Cryptochloris wintoni*), and therefore deserve strict protection.

Conservation of the Woodbrush and Agatha Forests
These two forests in north-eastern Transvaal are the only known localities for Gunning's golden mole (*Amblysomus gunningi*), and should therefore be conserved.

5.2.17 Tanzania

Conservation of Eastern Arc forests
These forests, especially the Usambaras and Ulugurus, are very important for rare species, including the black-and-rufous elephant-shrew (*Rhynchocyon petersi*), and seven species of shrew: *Crocidura monax* (Ulugurus and Kilimanjaro); *C. tansaniana* (Usambara); *C. telfordi* (Uluguru); *C. usambarae* (Usambara); an undescribed species of *Crocidura* (Mount Rungwe and the Uzungwa Mountains); *Sylvisorex howelli* (Usambara and Uluguru); and *Myosorex geata* (Uluguru). The current IUCN project aimed at conserving forest in the East Usambaras through an integrated development approach is very promising, and should serve as a model for other areas.

Conservation of coastal forests on the mainland and Zanzibar
Management policies in the Pugu Forest Reserve and other coastal forests in Tanzania should be reviewed in terms of protecting sufficient habitat for black-and-rufous elephant-shrews. High priority should be given to securing habitat for the subspecies on Zanzibar, *R.p. adersi*, notably the Muungwi Forest, Muyuni Coastal Strip, and Uzi Island. See Appendix 3 for more details.

5.2.18 Uganda

Conservation of forests in western Uganda
This region is very important for a number of threatened species: the Ruwenzori otter-shrew (*Micropotamogale ruwenzorii*); *Paracrocidura maxima*; *Ruwenzorisorex suncoides*; *Crocidura selina* (known only from the Mabira Forest near the shore of Lake Victoria); and outlying populations of the chequered elephant-shrew subspecies, *Rhynchocyon cirnei*

Grant's golden mole (*Eremitalpa granti*)
(Photo by G. C. Hickman)

stuhlmanni. These taxa are all shared with Zaire, and in the case of the two shrews, with Rwanda as well. Ongoing forest conservation measures should be supported and strengthened.

5.2.19 Zaire

Conservation of forests on the eastern border of Zaire
A large number of threatened species occur in this area, notably: the Ruwenzori otter-shrew (*Micropotamogale ruwenzorii*) (also in Uganda); *Crocidura kivuana* (Kahuzi Mountains: endemic); *C. lanosa* (also in Rwanda); *C. stenocephala* (Kahuzi Mountains: endemic); *Paracrocidura maxima* (also in Rwanda and Uganda); *P. graueri* (Itombwe Mountains: endemic); *Sylvisorex vulcanorum* (also in Rwanda); *Ruwenzorisorex suncoides* (also in Rwanda and Uganda); and *Myosorex schalleri* (Itombwe Mountains: endemic). The Virunga and Kahuzi-Biega National Parks are of particular importance, and the current projects aimed at strengthening their management need long-term support. Conservation measures should also be started in the Itombwe Mountains.

Conservation of the central rainforests
Forest conservation measures are needed to guarantee the survival of the following endemic shrews: *Crocidura congobelgica* (in the Ituri Forest); *C. crenata*; *C. lanosa*; *C. ludia*; *C. polia* and *Myosorex polli*.

5.2.20 Zambia

Conservation of gallery forests in north-western Zambia
The shrew *Crocidura ansellorum* is known only from gallery forests along the rivers of north-western Zambia, specifically the Kasombu Stream (Mwinilunga District) and the Nyansowe Stream (Solwezi District). The species is dependent upon the conservation of these habitats.

5.2.21 Zimbabwe

Conservation of the habitat of the yellow golden mole
The yellow golden mole (*Chlorotalpa obtusirostris*) occurs in alluvial soils, dry river beds and sand dunes in the south of the country, as well as in neighbouring South Africa and Mozambique; its habitat is probably in need of protection.

5.3 Field Surveys

The Insectivora and elephant-shrews are very poorly known compared with many other mammal species, as they are generally small and secretive. In many cases, it is difficult to formulate conservation priorities for these species because the basic distributional data are not available. Surveys of selected sites throughout the continent and on Madagascar are required to improve our understanding of distributional patterns and taxonomy, particularly for regionally endemic species with restricted or fragmented ranges. Where selected rare species are targeted, these surveys should be based on live trapping techniques wherever possible. Surveys need not necessarily focus solely on the Insectivora and elephant-shrews, but can include other species as well, so long as appropriate techniques are used for efficiently recording members of these taxa.

The following specific field surveys are recommended for the African Insectivora and elephant-shrews. This section is organised on a country by country basis.

5.3.1 Angola

General conservation assessment of the South African hedgehog
There is some concern about the status of this species, and assessment is needed to draw up conservation guidelines.

5.3.2 Botswana

General conservation assessment of the South African hedgehog
There is some concern about the status of this species, and assessment is needed to draw up conservation guidelines.

5.3.3 Cameroon

Surveys of forests in the west and south
The montane and lowland forests in southern and western Cameroon should be surveyed to determine the effects of deforestation on regional endemics, and to gather new distributional information on threatened species.

5.3.4 Central African Republic

Survey of the Boukoto area
This survey should assess the status of the shrew *Crocidura grassei* to determine whether it occurs more widely than is currently known.

5.3.5 Ethiopia

Survey of mountain forests
The Ethiopian Highlands comprise an area that is relatively rich in endemic shrew species. Small mammal surveys carried out by visiting expeditions, or as part of protected areas surveys, should include techniques appropriate for the Insectivora. Priority sites should include existing or proposed protected areas.

5.3.6 Gabon

Survey of the northern forests
The conservation requirements of the four rare species of shrew in this area should be assessed.

5.3.7 Guinea

Survey of Mount Nimba and surrounding regions
Surveys should be carried out to complement those already effected by Vogel (1983) on Mount Nimba, and focus on areas being affected by human settlements, forestry, and mining within the range of the Nimba otter-shrew (*Micropotamogale lamottei*) and of the shrew *Crocidura nimbae*. Live trapping techniques are strongly recommended, and surveys should seek to ascertain whether or not either of these species occurs more widely in the forested mountains in the south of the country.

5.3.8 Ivory Coast

Survey of Mount Nimba and surrounding regions
Surveys should be carried out to complement those already effected by Vogel (1983) on Mount Nimba, and focus on areas being affected by human settlements and forestry within the range of the Nimba otter-shrew (*Micropotamogale lamottei*) and of the shrew *Crocidura nimbae*. Live trapping techniques are strongly recommended, and surveys should seek to ascertain whether or not either of these species occurs more widely within Ivory Coast (perhaps in Tai National Park).

Survey for shrews in moist savanna habitats
The conservation needs of the endemic shrew *Crocidura wimmeri* around Adiopodoume should be assessed. Any remaining forested swamps should be protected.

5.3.9 Kenya

Survey of coastal forests
The status of the coastal forests in Kenya is being determined by the National Museums of Kenya Coast Forest Survey (S.A. Robertson, pers. comm.), funded by WWF. The results of this research should be invaluable in determining the extent and status of Insectivora and elephant-shrew habitat along the coast, which is under pressure from forestry practices and agriculture. However, a specific survey of the distribution and abundance of the forest-dwelling Insectivora and elephant-shrews is required to complement the existing botanical survey. As part of the elephant-shrew survey, it is especially important to develop a standard survey technique to determine the occurrence and relative abundance of elephant-shrews in an area. The status of the golden-rumped elephant-shrew (*Rhynchocyon chrysopygus*) needs to be assessed in forest patches north of Mombasa as far as the Boni Forest, and that of the black-and-rufous elephant-shrew (*R.p. petersi*) in isolated forest patches south of Mombasa. It is also unclear whether *Rhynchocyon* elephant-shrews occur in the dense scrub on coral rag along the coast. An assessment is also needed to determine whether these elephant-shrews can exist in highly modified habitats, such as fallow cashew plantations, or if they only use these areas when they border a larger expanse of undisturbed habitat. The impact of trapping and hunting the forest-dwelling elephant-shrews for food by local people along the Kenya coast should also be determined. This work might best be done in collaboration with an anthropolo-

The four-toed elephant-shrew (*Petrodromus tetradactylus*)
(Photo by G. B. Rathbun)

gist familiar with the coastal people. For the Insectivora it is important to determine the extent of local endemicity. More details are provided Appendix 3.

Survey of Taita and Chyulu Hills
The status of the isolated and severely fragmented patches of forest on the Taita Hills should be determined, along with the status of the four-toed elephant-shrew subspecies, *Petrodromus tetradactylus sangi*. A survey of the woodland and forest habitats between Kibwezi and the Chyulu Hills for the occurrence of the same subspecies should be undertaken.

Survey of Mount Gargues
A survey is needed of the montane forest and savannas of Mount Gargues to gain more information on the conservation needs of the endemic shrew *Crocidura raineyi*.

5.3.10 Lesotho

General conservation assessment of the South African hedgehog
There is some concern about the status of this species, and assessment is needed to draw up conservation guidelines.

Survey for Sclater's golden mole
A survey is needed to ascertain whether a significant population of this species survives in Lesotho, and if so, what conservation measures are needed.

5.3.11 Liberia

Survey of Mount Nimba and surrounding regions
Surveys should be carried out to complement those already effected by Vogel (1983) on Mount Nimba, and focus on areas being affected by human settlements, forestry, and mining within the range of the Nimba otter-shrew (*Micropotamogale lamottei*) and of the shrew *Crocidura nimbae*. Live trapping techniques are strongly recommended, and surveys should seek to ascertain whether or not either of these species occurs more widely in the forested areas of the country.

5.3.12 Madagascar

There is little baseline information on the distribution and abundance of most small mammals in Madagascar. Information is particularly sparse on the furred tenrecs. A series of selected surveys is required to identify critical areas and priority species. For some regions, trapping programmes within existing or proposed protected areas is appropriate, but larger blocks of natural forest as well as freshwater and marshland environments are not protected and require investigation. Surveys are required over a wide range of geographical localities and ecosystems, using trapping techniques that are showing some recent signs of being successful for some little-known species. In addition to snap trapping and live-trapping using standard "Sherman" and "National" traps, pitfall trapping can be highly productive. Pitfall traps have a low capture rate, but do reveal a range of smaller shrew-tenrecs (*Microgale* spp.) and mole-tenrecs (*Oryzorictes* spp.). For the aquatic tenrec (*Limnogale mergulus*), specialist live traps, such as eel traps, are required, which can be partially submerged in water.

For a single country, it may appear that the number and geographic diversity of surveys recommended below is unrealistic. However, there are several integrated conservation and development projects being developed in Madagascar, all of which include biological inventory and survey programmes, owing to the country's high conservation priority status. There are also a number of small mammal research projects being carried out that involve surveys at additional localities. In addition, the creation of a Biodiversity Planning Service is likely to occur by 1992. This unit will be responsible for a survey-based analysis of biodiversity patterns in Madagascar, and will include all of the above sites within its survey system.

Localities at which small mammal surveys are being implemented or planned at the moment include the Antsiranana area, Marojejy, Masoala, Mananara Nord, Andasibe, Ranomafana, Andohahela, south-eastern coastal forests, Beza Mahafaly, Morondava, Ambatovaky, and Manongarivo. This still leaves several gaps, particularly in the Western Floristic Domain, but some of these may be covered by encouraging further student projects and surveys, either with personnel from Madagascar or from other countries visiting for short-term expedition projects.

5.3.12.1 Eastern Rainforest

Surveys are required in a range of different rainforest ecosystems spanning the fragmented forest belt extending from Tolagnaro (Fort Dauphin) in the south-east to Marojejy in the north-east and crossing westwards to the Sambirano rainforests in the vicinity of Ambanja. It is equally important to carry out surveys in the forest isolates on the central plateau, including Ambohijanahary, Ambohitantely, and Kalambatritra, as well as the isolated forest of Montagne d'Ambre in the extreme north.

The following specific localities are recommended for surveys:

Survey of Sambirano forests
The Sambirano forests form a small distinctive mainland enclave in the north-west, extending to the island of Nosy Be. The largest undisturbed block lies on the plateau of Manongarivo which is a Special Reserve. This protected area is the most advisable survey locality.

Survey of Marojejy Massif
The Marojejy Massif is a Strict Nature Reserve (Réserve Naturelle Intégrale 12) situated between Andapa and Sambava in the north-east. This is the only intact floristic High Mountain Domain and spans an altitudinal range of 75-2,133 m. Earlier surveys indicate a high local endemicity among plant and reptile species.

Survey of Masoala Peninsula
The Masoala Peninsula lies in the north-east and has a high level of local plant and animal endemism. It is currently proposed as a national park area, and will be the subject of a wide range of biological inventories. The newly discovered dark pigmy shrew-tenrec (*Microgale pulla*), was recently described from Antongil Bay opposite the peninsula and might even occur at this site.

Survey of Ambatovaky Special Reserve
This reserve comprises the largest protected extent of Eastern Domain lowland rainforest. It has been surveyed in 1990 by the Madagascar Environmental Research Group (MERG), and two relatively little-known tenrecs, the striped shrew-tenrec (*Microgale melannorachis*) and the lesser long-tailed shrew-tenrec (*M. longicaudata*), were recorded.

Survey of Zahamena and the forests east of Lake Aloatra
The forests in this area are poorly known. Zahamena Strict Nature Reserve (R.N.I. 3) would provide an excellent locality for a survey, spanning an altitude of 100-1,500 m.

Survey of Anjazorobe
To the west of Lake Aloatra, a now almost isolated forest remnant extends from the main eastern forest block, south of Mandraka to the west of the lake. Anjazorobe is readily accessible from Antananarivo by road and is close to this forest extension, providing a means of examining a representative selection of this forest's fauna. It is currently being surveyed.

Survey of Ranomafana Est region
A large forest block near Ranomafana Est is currently being proposed as a national park. A range of rare bird and mammal species have been recently recorded there, including the aquatic tenrec (*Limnogale mergulus*). It is possible to survey forest ranging in altitude from 400-1,500 m. Biologists surveying this locality should be encouraged to extend their activities as far north as possible in order to cover the relatively unknown and unprotected fragmented forest chain that extends over a distance of 300 km. See Appendix 1 for more details.

Survey of Andringitra Massif
Surveys carried out 20 years ago in the Andringitra Strict Nature Reserve (R.N.I. 5) indicate a high small mammal species diversity, including some localized south-eastern rainforest species. This High Mountain Domain is accessible from Ambalavao and Ihosy, and spans an altitudinal range of 750-2,630 m, but natural forest extends to lower altitudes to the east of this protected area.

Survey of Midongy Atsimo (Midongy du Sud)
The rainforests in the region of Midongy Atsimo are probably the largest in southern Madagascar. They are probably contiguous with those of Andohahela in the extreme south-east and Andringitra to the north-west. Surveys in the Midongy-Atsimo region are likely to provide good indications of tenrec species composition in the south-east.

Survey of Andohahela
Andohahela lies at the extreme southern limit of the eastern rainforests, and includes transitional zones between these formations and the southern spiny semi-arid formations. It is a Strict Nature Reserve (R.N.I. 11), and is considered a priority protected area. In addition to those already conducted, another series of biological inventories is planned.

Survey of east coast rainforest isolates
Several coastal rainforest isolates occur, particularly in the south-east. These include Manombo Special Reserve, and Mandena and Saint Luce Forests. Mandena and Saint Luce were surveyed in 1989.

Survey of upland rainforest isolates
There are several isolated rainforest blocks on the central plateau and to the north of the main eastern rainforest block. It would be useful to conduct small mammal surveys in Ambohitantely Special Reserve 130 km north of Antananarivo, at Ambohijanahary Special Reserve 300 km west of Antananarivo on the western edge of the central plateau, and in Montagne d'Ambre National Park in the extreme north. Two shrew-tenrec species, the northern lesser long-tailed shrew-tenrec (*Microgale prolixacaudata*) and Drouhard's shrew-tenrec (*M. drouhardi*), are recorded from the extreme north, and are generally considered to have originated from Montagne d'Ambre. Their taxonomic status is unclear at present, and surveys would provide clarification and thus realistic range limits to these apparently localised species (MacPhee 1987).

5.3.12.2 Western and Southern Ecosystems

Furred tenrecs are apparently poorly represented and found in only a few localities within the western and southern areas. However, it is not clear whether this apparent lack of diversity is real or simply the result of lack of surveys and museum collecting. Further surveys are therefore needed to resolve these questions. Survey localities should be restricted to natural forest areas, which are now highly fragmented within the Western Domain, and wetlands.

Survey of Andohahela
Parcelle 2 of Andohahela lies within the semi-arid Southern Domain close to this domain's eastern limits. This area is probably representative of assemblages of Insectivora within the eastern sector of the south, and is close to sub-fossil localities of *Microgale decaryi*.

Survey of Tsimanampetsotsa
This strict nature reserve (R.N.I. 10) lies in the south-west, and supports spiny thicket dominated by endemic Didieriaceae on limestone karst, and sand dune/alkaline lake fringe vegetation. It is one of the few shrew-tenrec localities outside the Eastern Centre of Endemism, where *Microgale pusilla* has been collected recently (MacPhee 1987).

Survey of Lac Ihotry
The moist habitats of this lake fringed by western deciduous forest and spiny thicket would be probable collecting sites for scattered pockets of furred tenrec species.

Survey of Kirindy Forest
There are currently no protected areas between Tsinamampetsotsa south of Toliara and Andranomena, north of Morondava, a distance of 400 km. As at Lac Ihotry, surveys of Kirindy Forest would provide information on possible tenrec assemblages in this area, as well as supplying information on other faunal groups.

Survey of Zombitse and Vohibasia Forests
These two forests are isolated pockets of western deciduous forest in the south-west. They are unusual within this region in that the greater hedgehog-tenrec (*Setifer setosus*), while normally absent in the south-west, occurs at this site. The large-eared tenrec (*Geogale aurita*) is also present.

Survey of Forest concession Centre de la Formation Professionelle Forestière, Morondava
This natural forest north of Morondava is currently exposed to low-level logging and has excellent facilities and access. It has been surveyed annually since 1988, producing shrew-tenrec specimens referable to either the short-tailed shrew-tenrec (*Microgale brevicaudata*) or the western short-tailed shrew-tenrec (*M. occidentalis*), and one specimen resembling the lesser long-tailed shrew-tenrec (*M. longicaudata*).

Survey of Bemamba and Masama Lakes, and Tsingy de Bemaraha
A series of freshwater and alkaline lakes occur within the western deciduous forest of Tsimembo on the coastal plains to the west of the Tsingy de Bemaraha Strict Nature Reserve (R.N.I. 9). These may provide suitable sites for oryzorictine tenrecs, particularly as the western short-tailed shrew-tenrec (*M. occidentalis*) was collected only some 120 km to the north. The Tsingy de Bemaraha plateau is relatively moist in comparison to surrounding areas and may also support furred tenrecs. This region is known for its isolated population of the rodent *Nesomys rufus*, a species characteristic of the eastern rainforests. This western population, *N.r. lambertoni*, might in fact be a full species, distinct from its eastern relatives.

Survey of north-western forests
Several deciduous forest blocks on or near the north-western coast include the Tsingy de Namoroka Strict Nature Reserve (R.N.I. 8), and Bamarivo, Maningozo, and Kasijy Special Reserves. Information on these is limited to Namoroko, where a few brief surveys have been conducted. It would be useful for developing a priority protected areas rating to carry out surveys, including inventories, of small mammals at these sites.

Survey of the extreme north
The area surrounding the city of Antsiranana in the extreme north comprises a range of deciduous forest formations on limestone karst, and the Montagne d'Ambre rainforests. While it is generally held that the two localized species, the northern lesser long-tailed shrew-tenrec (*Microgale prolixacaudata*) and Drouhard's shrew-tenrec (*M. drouhardi*), originated from Montagne d'Ambre, the possibility remains that they came from dry forest localities, especially as sub-fossil remains of the eastern rainforest species *M. talazaci* have been recorded in cave deposits in dry forest. The taxonomic status of the two extant isolated species is unclear, and surveys would help clarify their position and thus their apparent range.

5.3.13 Malawi

Survey the area around Livingstonia
The isolated montane forest along the west side of Lake Malawi just north of the town of Livingstonia, which is the only known location for the chequered elephant-shrew subspecies, *R.c. hendersoni* needs to be surveyed, as neither the extent nor current status of this forest is known. The status of the subspecies in this area is also unknown.

Survey of northern and southern forests
A brief survey of habitat and distribution of the chequered elephant-shrew subspecies, *R.c. shirensis* and *R.c. reichardi* should be undertaken. In Malawi the forests and woodlands associated with the Mafinga Mountains and Nyika Plateau should be assessed for *R.c. reichardi*. Likewise Mounts Mulanje, Thyolo and other areas in the south should be surveyed for *R.c. shirensis*.

5.3.14 Mozambique

Survey of Quelimane area
A survey is needed in the Quelimane area, north of the mouth of the Zambezi River, to determine the type of habitat, its extent, and the status of the chequered elephant-shrew subspecies, *R.c. cirnei*, as this is the site of the type (and only) specimen.

General conservation assessment of the South African hedgehog
There is some concern about the status of this species, and assessment is needed to draw up conservation guidelines.

Survey for the yellow golden mole
A survey is needed for this species in the south of the country to assess status, distribution and conservation requirements.

5.3.15 Namibia

General conservation assessment of the South African hedgehog
There is some concern about the status of this species, and assessment is needed to draw up conservation guidelines.

Survey sand dune formations in Namibia
Surveys in these areas should be carried out to determine the status and conservation requirements of Grant's golden mole (*Eremitalpa granti*) in this habitat in the south of the country. See Appendix 2 for more details.

5.3.16 Nigeria

Survey of swamps near Kainji Lake National Park
This survey should assess the distribution of *Crocidura longipes*, in particular whether or not it occurs in the national park.

5.3.17 Rwanda

Survey of mountain forest species
These surveys should concentrate on the endemic shrews in the Nyungwe Forest and the Volcanoes National Park.

5.3.18 São Tomé and Príncipe

Survey for the endemic shrew species
A survey is needed to assess the status and conservation requirements of the very rare *Crocidura thomensis*.

5.3.19 Somalia

Survey of the Boni Forest
The Boni Forest extends across the Kenya-Somalia border. It is possible that the golden-rumped elephant-shrew (*Rhynchocyon chrysopygus*) occurs in this habitat, but this needs to be confirmed.

Survey of the Giohar region
A survey is needed of the Giohar region of Somalia to obtain information on the current status of the golden mole (*Amblysomus tytonis*). This species is known only from the type specimen collected in this area and it occurs well beyond the range of other members of the family.

5.3.20 South Africa

General conservation assessment of the South African hedgehog
There is some concern about the status of this species, and assessment is needed to draw up conservation guidelines.

Survey of the Gouna area
A survey is needed of the Gouna area to relocate Visagei's golden mole (*Chrysochloris visagiei*), which is known from just one specimen from this site.

Survey of the sand dunes at Port Nolloth
This is the only known site for De Winton's golden mole; its conservation needs should be assessed, and surveys should be made for additional populations.

Survey of Lambert's Bay
Van Zyl's golden mole is only known from this site; its conservation needs should be assessed, and surveys should be made for additional populations.

Survey sand dune formations in south-western South Africa
Surveys along the coast should be carried out to determine the status and conservation requirements of Grant's golden mole (*Eremitalpa granti*) in this habitat. See Appendix 2 for more details.

Surveys in coastal southern and south-eastern South Africa
A range of surveys is required to establish the taxonomic status (where appropriate), and distribution, abundance and conservation requirements of the little-studied golden moles and shrews that are restricted to this region.

General golden mole surveys
There are a number of other rare golden mole species in South Africa, for which surveys are needed and conservation recommendations made. These include Sclater's golden mole (*Chlorotalpa sclateri*), the yellow golden mole (*Chlorotalpa obtusirostris*), Gunning's golden mole (*Amblysomus gunningi*), Juliana's golden mole (*Amblysomus julianae*) and the rough-haired golden mole (*Chrysospalax villosus*). See Appendix 2 for more details.

Survey for rare shrews
This should concentrate on the very poorly known *Crocidura maquassiensis* and *Myosorex longicaudatus*, and should assess their conservation requirements.

5.3.21 Swaziland

Survey for rare shrews
This should concentrate on the very poorly known *Crocidura maquassiensis*, and should assess its conservation requirements.

5.3.22 Tanzania

Survey of coastal forests
The status of the black-and-rufous elephant-shrew (*R.p. petersi*) and its habitat along the coast in northern Tanzania is unknown. The Pugu Forest, south-west of Dar-es-Salaam, has been greatly reduced in area recently (Howell 1981); the status of the black-and-rufous elephant-shrew should be determined in the Pugu Forest and other less well known areas along the Tanzanian coast. See Appendix 3 for more details.

Survey of Zanzibar forests
The distribution and status of the black-and-rufous elephant-shrew subspecies, *R.p. adersi* and its forest habitat on Zanzibar and Mafia islands should be determined.

Survey of Eastern Arc forests
The black-and-rufous elephant-shrew (*R.p. petersi*) is known from several isolated Eastern Arc Forests, such as those on the Nguru, Uluguru, and Usambara Mountains (Corbet and Hanks 1968; Kingdon 1974). Like many of the forested areas of eastern Africa, these montane areas have been affected by tree-harvesting and clear-cut and burn agriculture. Forest destruction is especially critical in the East Usambaras, and to a lesser extent the Ulugurus (Stuart 1981). The current status of both the elephant-shrew and its habitat should be determined in all of these forests. It is not clear whether any *R.p. petersi* occur in the Pare Mountains, which should be assessed. The other Eastern Arc Forests (Ukaguru, Rubeho, Uzungwa, and Mahenge) probably support the chequered elephant-shrew subspecies, *R.c. reichardi*. However, these isolated forests should all be surveyed. Part of this survey should include an assessment of the effects that hunting elephant-shrews for food has on the populations, and whether elephant-shrews are occupying altered habitats, such as tea estates and residential gardens. Observations of animals in these habitats may be the result of wanderers from adjacent forests. The surveys should also aim to identify the conservation needs of the shrews, in particular the endemic species in the Usambaras and Ulugurus.

Survey of Mount Meru
A specimen of the four-toed elephant-shrew subspecies, *Petrodromus tetradactylus sangi* in the British Museum (Natural History) is labelled "Kibwezi", which could either be the site in Kenya (see section 5.3.9 Kenya), or the town of the same name in the Mount Meru area. A survey of the forests on Mount Meru is needed to clarify this situation.

Survey of forests in southern and western Tanzania
A brief survey of habitat and distribution of the chequered elephant-shrew subspecies, *R.c. reichardi* should be undertaken in isolated forests on mountains and plateaus near lakes Tanganyika and Malawi. There is little information on its current status. In Tanzania, likely sites are Mount Rungwe, the Poroto Mountains, the Kipengere Mountains, the Matengo Highlands, the Gombe Stream National Park, the Mahale Mountains National Park, and the Karema and Mbizi forests. These surveys should include an assessment of the newly discovered shrew from Mount Rungwe and the Uzungwa Mountains, southern Tanzania.

5.3.23 Uganda

Survey of isolated lowland forests
Bwamba (Semliki), Bugoma, Budongo, and Mabira forests should be surveyed for the chequered elephant-shrew subspecies, *R.c. stuhlmanni*, which might be at risk locally. In addition, it is not known whether this subspecies occurs in the forests south of Bwamba (in Itwara, southern Kibale, and the Kasyoha-Kitomi/Kalinzu/Maramagambo complex). The altitude of these forests may be too high, but they still should be surveyed. The survey should include an assessment of the status of the endemic shrew *Crocidura selina*.

5.3.24 Zaire

Surveys for rare Insectivora in the east of the country
Nine species of conservation concern occur in this area, including the Ruwenzori otter-shrew (*Micropotamogale ruwenzorii*). The distributions and conservation requirements of all these species need further assessment, and priority sites are the Kahuzi, Itombwe and Virunga Mountains.

Surveys for endemic shrews in the lowland rainforests
Three endemic shrew species occur in this region, and their conservation requirements need to be assessed. Priority survey sites are the Ituri Forest and the Maiko, Salonga and Kahuzi-Biega National Parks.

5.3.25 Zambia

Survey of north-eastern highlands
A brief survey of the habitat and distribution of the chequered elephant-shrew subspecies, *R.c. reichardi* should be undertaken in isolated forests on the Makutu Mountains.

Survey of north-western gallery forests
This survey should search for the endemic shrew *Crocidura ansellorum* in Mwinilunga and Solwezi Districts.

5.3.26 Zimbabwe

General conservation assessment of the South African hedgehog
There is some concern about the status of this species, and assessment is needed to draw up conservation guidelines.

Survey for the yellow golden mole
A survey is needed for this species in the south of the country to assess status, distribution and conservation requirements.

Survey for rare shrews
This should concentrate on the very poorly known *Crocidura maquassiensis*, and should assess its conservation requirements.

5.4 Research

Research programmes are neded to determine the habitat requirements of rarer species, the factors affecting distribution and survival, and to provide baseline data for any necessary captive management programmes. Such research may be carried out by both resident and visiting expatriate biologists, and can serve as useful training exercises for local students in the form of M.Sc. or Ph.D. projects. Foreign research institutions should provide training, and financial and material support to African counterparts, who often face severe limitations on conducting such activities.

The following broad-scale activities are recommended:

5.4.1 Systematics

Based on material collected during the above surveys (Section 5.3), reviews of the taxonomic status of problem species are required, to determine the validity of the taxa and hence their real conservation status.

For example, many of the furred tenrecs have been recorded only once or on a few occasions, and are furthermore known only from early collecting expeditions, sometimes with poor collecting locality information. There is taxonomic confusion surrounding a number of species. MacPhee (1987) has produced the most recent review, reducing the genus *Microgale* from 19 extant forms to 11. While this is a helpful clarification of *Microgale* systematics, there is some concern that this might represent an over-simplification, notably by sinking too many species into the *M. cowani* complex. MacPhee (1987) recognised the need for studies on the locomotion of some *Microgale* species to clarify differences between disputed taxonomic forms. *Microgale* species are apparently maintained relatively easily in captivity, and so such studies could be incorporated into student projects. Other examples of groups in need of much more taxonomic research include the African shrews (in particular the genus *Crocidura*) and the golden moles. It must be emphasised that the current trend away from basic taxonomic research is extremely alarming from a conservation perspective, and governments throughout the world are urged to provide increased support to museums and other scientific institutions engaged in taxonomic work.

5.4.2 Ecology

Further research is required on the ecological requirements of selected species. Such projects are required most urgently in countries where little research on the Insectivora and elephant-shrews has been carried out, and may be best effected by encouraging and assisting in establishing individual student short-term projects. Also needed is research on the factors influencing species diversity and the co-existence of large numbers of species in forest environments. Research should also focus on the tolerance of threatened species to forest exploitation and fragmentation, and such studies should generally be carried out in centres of species diversity and endemism.

Priority groups for ecological research include:

Golden moles
Research on the habits and requirements of threatened golden mole species is a high priority, since it is clear that many of these species have very particular, yet obscure needs. See Appendix 2 for more details.

Elephant-shrews
Although well known taxonomically, there are still large gaps in our ecological knowledge of elephant-shrews, in particular habitat requirements, reactions to habitat disturbance, and breeding ecology, and these topics should be addressed through a series of discrete field projects. It is also a high priority to develop an efficient, reliable method of surveying for the forest-dwelling elephant-shrews (*Rhynchocyon* and *Petrodromus*), so that the numerous status surveys proposed in this plan (section 5.3) can be initiated. See Appendix 3 for more details.

Madagascar Tenrecs
MacPhee (1987) has suggested that Madagascar tenrecs might not be a conservation risk group, based on the wide distribution of the family, and the fact that some species occur in a wide variety of habitats, including man-made environments. Studies by Nicoll *et al.* (1988) contest this, showing that not all tenrec species are able to survive in man-made environments, and that forest species might be more conservative ecologically. Nevertheless, more research is needed on the degree to which threatened tenrec species are able to adapt to habitat modification.

5.5 Captive Breeding

Captive breeding of the African Insectivora and elephant-shrews is still in its infancy, partly because so little is known about the breeding biology and ecological requirements of these species. Captive breeding can never be a substitute for the conservation of a species in the wild through the protection of sufficient areas of habitat to support viable populations, but it can be a useful adjunct, especially for seriously threatened species with very limited ranges. The research which can be carried out on species in captivity can provide very useful information to assist *in situ* conservation efforts. Captive propagation need not necessarily be carried out by zoological gardens, but may also be effected in some cases by university departments with experience in Insectivora or elephant-shrew captive management, either on the African continent or elsewhere. Surveys and field programmes should always endeavour to document information which could have relevance to future captive propagation programmes.

Priority threatened species include:

Golden moles
Recommended target species include all species with a threatened species status, particularly those considered to be Rare or Vulnerable. The giant golden mole (*Chrysospalax trevelyani*) is suggested as the most suitable species to start with as it eats a wide variety of food, and might therefore be less difficult to maintain in captivity. Copulation and birth have been

observed in this species, and much could be learned about the golden moles in general from this species (partly because it forages on the surface, despite its obvious digging specialisations).

Otter-shrews

Captive propagation techniques should be determined for the two *Micropotamogale* (lesser otter-shrew) species in semi-captive conditions in the wild, coupled with an assessment of the need to establish a captive colony. Vogel (1983) has already provided a considerable quantity of baseline information for the Nimba otter-shrew (*M. lamottei*), and a captive breeding programme for this very threatened species is now a very high priority.

Madagascan tenrecs

The aquatic tenrec (*Limnogale mergulus*) is the highest priority for a captive propagation programme (see Appendix 1 for further information). This possibility is currently being explored by a North American zoo. There are also other opportunities for an integrated programme to determine captive breeding requirements and techniques, based on species which are already held in captivity. The best-known species in this regard are non-threatened spiny tenrecs, all of which are currently in zoological collections and have been bred in captivity. From available information, only three species, the tail-less tenrec (*Tenrec ecaudatus*), the greater hedgehog-tenrec (*Setifer setosus*), and the lesser hedgehog-tenrec (*Echinops telfairi*), are held in collections outside Madagascar. Several furred tenrecs, including the large-eared tenrec (*Geogale aurita aurita*) and a few shrew-tenrecs (*Microgale* sp.) are in the Parc Botanique et Zoologique de Tsimabazaza in Madagascar, the island's national zoo. This collection will be expanded over the next two years, but it is recommended that populations be established in other zoological collections elsewhere that have a genuine interest in conservation and which are prepared to enter into collaborative agreements with Tsimbazaza. All furred tenrec species are recommended for captive breeding, in addition to the streaked tenrec (*Hemicentetes nigriceps*), which has the most restricted range of any spiny tenrec species.

The black-headed streaked tenrec (*Hemicentetes nigriceps*)
(Photo by M. E. Nicoll/BIOS)

Elephant-shrews

A renewed attempt to care for *Petrodromus* and *Rhynchocyon* species in captivity should be initiated, using modern animal management practices and incorporating the best information available on their biology. In this regard, it important that the Frankfurt Zoo be encouraged to document their successful care of the golden-rumped elephant-shrew (*R. chrysopygus*). The first efforts to breed these genera should be made with species that are common. Because large areas of habitat are destroyed every year, individual animals for these programmes should be acquired in areas destined for destruction. Once captive breeding methods are successfully developed for these more common "model" species, then consideration should be given to establishing captive breeding programmes for the most threatened species, with the ultimate aim of reintroducing the animals as soon as habitat can be secured.

5.6 Priorities for Action

At the present time, in view of the very limited information available on most of the species, it is not possible to develop a clear ranking of priorities among all the activities listed in this Action Plan. However, this has been done for those projects that relate to the elephant-shrews.

The proposed projects related to elephant-shrews have been ranked according to a subjective appraisal of the following factors, listed in decreasing importance: a) status of habitat; b) status of the elephant-shrews; c) distinctiveness of elephant-shrews (taxonomic status and extent of distribution); d) ease of accomplishing the task (logistical and financial considerations); and e) a subjective assessment of the task's importance in maintaining biotic diversity. Taking these factors into account, the ranking of elephant-shrew projects is as follows (in decreasing priority order):

- Development of survey techniques
- Survey of coastal forests in Kenya
- Survey of Taita and Chyulu hills in Kenya
- Survey of coastal forests in Tanzania
- Survey of Zanzibar forests
- Conservation of the Taita Hills in Kenya
- Survey of Eastern Arc forests in Tanzania
- Conservation of coastal forests in Tanzania
- Conservation of coastal forests in Kenya
- Conservation of Eastern Arc forests in Tanzania
- Survey of Livingstonia area, Malawi
- Survey of Quelimane area, Mozambique
- Survey of Mount Meru in Tanzania
- Survey of isolated lowland forests in Uganda
- Survey of northern and southern forests in Malawi
- Survey of Boni Forest in Somalia
- Captive breeding research

Appendix 1: An Evaluation of the Status of the Aquatic Tenrec *Limnogale mergulus*

The Conservation Need

The aquatic, or web-footed tenrec, *Limnogale mergulus*, is the only aquatic member of the family Tenrecidae in Madagascar, is one of its least known members, and is almost certainly the most severely threatened species of this remarkable group of animals. The exact range distribution of the aquatic tenrec is unknown, but is almost certainly confined to the eastern section of the country, where it is known by local people who happen to catch them in fish traps. As regards habitat choice, the aquatic tenrec is restricted to small, fast-flowing streams, where they are believed to feed on freshwater crayfish, aquatic insect larvae and small crustacea.

In Madagascar, fast-flowing rivers where suitable aquatic plants are abundant are becoming increasingly isolated as slash-and-burn ('tavy') agriculture spreads, fragmenting the remaining vestiges of escarpment or montane plateau forest. It has also been suggested that the aquatic tenrec is associated with the aquatic plants, *Aponogeton fenestralis* and *Hydrostachis madagascariensis*, as these provide a rich supply of invertebrates within their tangled root systems. *A. fenestralis* is frequently collected as an aquarium ornamental, and has suffered local extirpation as a result. It appears, therefore, that the aquatic tenrec, which is naturally patchily distributed and, if similar to its aquatic otter-shrew relatives in west and central Africa, found only at low densities, is facing an immediate threat through reduction of suitable environments. It may also face difficulties in traversing the anthropogenic environments separating these diminishing localities. It is essential that the remaining aquatic tenrec localities are identified and that the correct decisions are taken concerning the species immediate and long-term survival.

As recently as the 1960s, this species was known from just a few localities on the eastern edge of the central high plateau and the eastern escarpment. Four of the five localities were fast-flowing montane streams with a profusion of the aquatic plants *A. fenestralis* and *H. madagascariensis*. The fifth was a lake. M.E. Nicoll visited two of these sites in 1986/1988 as part of a countrywide WWF-based survey of protected areas and biologically important sites. In these two areas, Vohemar River in the central eastern region and the Akaratra Massif, major habitat changes and river sedimentation were found to be taking place. No feeding signs were observed and the species is now thought to be extremely rare at these sites. During the same survey period, three additional sites were identified where, according to local people, the aquatic tenrec is still present and is often captured accidentally in their fish traps. These sites are: the region of the proposed Ranomafana National Park; immediately northeast of the Andringitra Strict Nature Reserve (Réserve Naturelle Intégrale 5); and 35 km south of Antsirabe. Of the many sites which were visited during the survey and where conditions appeared to be favourable, only the latter had abundant feeding signs of the tenrec.

Thus the limited information available on this species, as well as the ever-increasing threat to its habitat, strongly suggest that the aquatic tenrec may be a prime insectivore candidate for a captive breeding programme that would eventually be linked to re-introductions that could ensure survival of wild populations in protected areas. Whether this requires a small-scale trial programme to establish management techniques for this potentially difficult species, or a major programme to establish sufficient genetic diversity in captivity, has yet to be decided.

The aquatic tenrec (*Limnogale mergulus*)
(Photo by M. E. Nicoll/BIOS)

The Conservation Requirements

This research programme is primarily concerned with investigating the ecological requirements and conservation status of the aquatic tenrec. Its main purpose is to: (i) determine the current conservation status and requirements of what appears to be one of Madagascar's rarest mammals occupying a unique ecological niche on the island; (ii) evaluate the current threats to the central high plateau and eastern escarpment forests, of which the former has probably suffered more than any other region from natural habitat destruction; and (iii) evaluate the need for captive propagation to ensure the aquatic tenrec's survival.

1. Conservation Status and Requirements of the Aquatic Tenrec

The aquatic tenrec is one of only four aquatic representatives of one of the earliest placental mammal radiations, and is also likely to be a key carnivore in a patchily distributed and increasingly rare Malagasy river ecosystem. This project will be the first attempt to identify remaining populations of the aquatic tenrec, its abundance, ecological requirements, and the factors which influence its future survival. All available evidence, albeit very limited, points to this tenrec being one of the highest insectivore species survival priorities either on a worldwide basis or within Madagascar.

A survey component of this project examining a range of high plateau and eastern escarpment localities to identify aquatic tenrec sites will be important in determining the environmental correlates of the tenrec's presence. This survey should also identify the influence of changes in the natural environment on the tenrec's survival. Specific attention will be paid to the effects of deforestation resulting in higher soil erosion and siltation. A questionnaire will be devised to gather information from villagers in an attempt to determine whether, according to their knowledge, the distribution of this species has declined or not.

2. Current Threat to Escarpment and Eastern Central Plateau Rivers and Forests

The evaluation of known and potential aquatic tenrec sites will identify central plateau and eastern escarpment localities that remain sufficiently intact to contribute to a nationwide conservation policy to maximise biological diversity. For example, the Ankaratra Massif is currently unprotected, even though it still supports one of the few remaining plateau forests. Similarly, identification of the aquatic tenrec sites within the Ranomafana region will help define the most appropriate limits of the new protected area currently being planned.

3. Evaluating the Need for Captive Propagation of the Aquatic Tenrec.

Since it was first described, it has been clear that the aquatic tenrec is not abundant. Now that its natural environment is increasingly threatened, it is not clear if the wild population will remain viable in the future. One of the main objectives of this project will be to determine whether a captive propagation programme is required. Specifically, this will involve holding individuals in semi-captivity to determine husbandry and propagation techniques.

Background Information

The aquatic tenrec has been collected or recorded from just a few scatted localities. In 1965, Malzy became the first person to succeed in catching an aquatic tenrec alive at the Station Forestiere d'Apansandrano (19°37S, 47°04'E) on the Ankaratra Massif. Shortly thereafter, Gould and Eisenberg (1966) reported that their signs at this same site were relatively abundant, if highly localised. In later years, Gould and Eisenberg searched extensively in a region where Grandidier has previously collected the animal, on the Vohemar River at Ambodivoangy (18°46'S, 48°45'E). No animals were recorded, however, and they deduced that the animal was rare and highly localised. Since then, there were no further records, until the WWF - Protected Areas Programme survey found three additional sites in 1986-1988.

At one site on the eastern escarpment of the central plateau near Ranomafana Est (21°15'S, 47°28'E), signs of the aquatic tenrec occur sporadically in small fast-flowing streams and the animal is occasionally captured in fish traps. These aquatic tenrec localities should be included in the planned Ranomafana National Park. The second site (22°02'S, 46°55'E) lies 10 km north of the northern boundary of Réserve Naturelle Intégrale 5, Andringitra, in fast-flowing streams originating from the reserve. While villagers living immediately north of the reserve do not know of the animal's occurrence in the protected area, it is possible that isolated populations occur there. In the time available for survey, none were found in the reserve. The third site is near the village of Antalava, 35 km south of Antsirabe.

More recently, in 1989, R. D. Stone and E. Gould succeeded in capturing five animals at a site near Ranomafana, using techniques previously devised for the Pyrenean desman (*Galemys pyrenaicus*). Several of the animals were held in captivity for short periods of time and one individual was brought to Parc Tsimbazaza, where it was held for three weeks before being returned to the original trapping site. First-hand observations of these animals have already furnished baseline data on trapping and holding techniques.

Objectives

The main objectives of this project will be to:

- verify known localities and identify the physical and ecological characteristics of the aquatic tenrec sites;

- determine the ecological requirements - diet, spatial needs, etc., of the aquatic tenrec;

- determine the current threat to the aquatic tenrec through habitat changes, and promote conservation action where necessary and feasible;

- determine whether captive propagation is required and, if so, conduct a series of semi-captive trials to determine and evaluate captive breeding techniques.

Methods

1. Capture - marking studies at a known locality

Brief attempts to capture animals, using techniques established for the Pyrenean desman (*Galemys pyrenaicus*) and the giant otter shrew (*Potamogale velox*), will be carried out initially at the Ranomafana, Andringitra, Antsirabe, and Ampansandrano sites to select a workable population. Having selected a site, an intensive trapping programme will be carried out to estimate population density along particular streams. This will be followed up by an assessment of aquatic tenrec signs (faeces, food remains) to determine the extent of the species presence in the area. All sites will be analysed in terms of habitat type.

2. Brief survey of potential the aquatic tenrec sites

Following the development of appropriate surveying techniques, rapid surveys will be conducted at several sites on the eastern edge of the central plateau and the eastern escarpment. Suggested sites are in the Fianarantsoa region, the Antsirabe region, and near Andekaleka on the Antananarivo - Toamasina railway line.

3. Semi-captive trials

At the main study area, a small group of tenrecs (2-4) will be held in cages built on stream banks. Observations will be conducted to determine feeding regimes, activity patterns and nesting requirements. Cage design will be carefully planned in order to ensure adequate levels of activity and stimulate good fur condition - an important factor for all aquatic insectivores.

Local Involvement

This project will require considerable involvement with local communities in areas where the aquatic tenrec might occur. Involvement will stimulate discussion on environmental conservation issues at the village level, and require direct participation of local guides and assistants. In recent conservation projects, this has led to increased receptivity to conservation issues by villagers, and a better understanding of the broader conservation issues under investigation by the researchers. For this project, such participation would be particularly useful in the Andringitra and Ranomafana regions.

This project will involve personnel from the Parc Botanique et Zoologique de Tsimbazaza, the principal captive animal conservation facility in Madagascar. Tsimbazaza has recent experience with seven tenrec species and is the only captive facility to have bred the large-eared tenrec (*Geogale aurita*). In addition, the curator responsible for non-primate mammals has already worked extensively on the tenrecs and is therefore familiar with the techniques necessary for this project. Also, if the project is to be conducted by an expatriate it would be appropriate for training to be provided to a Malagasy student from the Université de Madagascar who could then use the data for a higher degree.

Conservation and Environmental Education

This project will provide valuable training for participating Tsimbazaza personnel in dealing with a species that is rare and that requires complex maintenance techniques. Such participation will also be a useful contribution to an expanding small mammal field expertise in this captive facility.

Ultimately, if this project leads to a captive propagation programme in a European or American zoological park, this captive facility will gain valuable experience that may be useful elsewhere in establishing future zoo populations of other threatened aquatic insectivores such as the Nimba otter-shrew (*Micropotamogale lamottei*), Ruwenzori otter-shew (*M. ruwenzori*), and Pyrenean desman (*Galemys pyrenaicus*).

Equally important is the enhanced awareness of the importance of the role that local inhabitants take in protecting Madagascar's unique wildlife and environment. This will be a major objective of the project participants.

Appendix 2: Conservation of the Golden Moles, Especially the Genus *Chrysospalax*

Background

Conservation of moles may seem enigmatic to many people. "Moles" (insectivorous subterranean mammals) in Europe have long been persecuted as familiar garden and pasture pests, and have been well-known characters in children's tales. Nonetheless, the African golden moles include many unique species, some of which are rare and endangered, and others about which we know absolutely nothing.

The distribution of most golden mole species is either highly restricted or patchy, or is very poorly known and difficult to define. For example, although the range of the golden moles is often given as extending throughout Angola, they are in reality known from just a single specimen in this country. Similarly, the Somali golden mole (*Chlorotalpa tytonis*) from Somalia is known only from owl pellets. Other forms of golden moles are restricted to isolated mountains. Compounding the study of the distribution of the golden moles are problems concerning the basic taxonomy of the group.

Objectives

The following project is considered necessary, since broad faunal surveys, including those for other insectivore groups, do not normally collect data on moles. Its objectives are:

1. To explore the potential for discovering new species which might otherwise disappear before their existence is known.

2. To collect specimens of little-known species which will assist in deciphering the taxonomy of the group.

3. To support studies of little known species, such as the Zulu golden mole (*Amblysomus iris*), to gain a more comprehensive picture of golden mole biology (which is probably very different from the biology of the better known moles of Europe, Asia and North America).

4. To encourage the conservation of areas vital for the survival of particular golden mole species.

5. To publicise the uniqueness and importance of golden moles to conservation authorities and the general public.

6. To explore the possibility of captive breeding programmes for species whose habitat is rapidly declining, such as the giant golden mole (*Chrysospalax trevelyani*).

Justification

A third of the known golden mole species are known from less than seven specimens each. This underlines how difficult it is to assess the conservation needs for these species. Other species, such as the Zulu golden mole, which is not particularly rare, is still very poorly known from a biological and ecological perspective. The broader our understanding of golden moles in general, the better the chance of developing effective conservation programmes for particular species of concern.

The conservation needs of a few species are, however, particularly prominent: the giant golden mole: the rough-haired golden mole (*Chrysospalex villosus*); and the Namib Desert golden mole (*Eremitalpa granti*). These can be summarised as follows:

The giant golden mole is known from less than 100 specimens in museums from around the world. Its large size (500+ g) and habit of foraging in the surface leaf litter (despite its blindness and other obvious digging adaptations) are unique. The forests of the eastern Cape represent its world distribution, and this habitat is being quickly reduced, fragmented, and degraded.

The rough-haired golden mole does not require forest, and hence it is much more difficult to pinpoint the grassland areas in which this species occurs. It would be highly desirable to locate at least one locality where this species could be reliably captured, so that a study could be undertaken on its habits. It is known from Transvaal, Natal and eastern Cape Province.

The Namib desert golden mole was little known until a doctoral thesis on the species was completed in 1989. This unique insectivorous mole lives in the sand dunes of the Namib Desert without producing permanent foraging tunnels. There still remain many interesting questions regarding its ecology, in particular: where are the young born?

Activities

The principal focus of future activities will be on faunal surveys, which will use the special traps and techniques required for the successful capture of golden moles. Whole carcasses should be preserved to maximize the information gathered from these animals. If possible, chromosome preparations should be made, in addition to the preparation of certain tissues for DNA analysis to assist in the resolution of taxonomic problems.

Student studies will be encouraged on all species; there is all too little information, even for the common species. These studies will include surveys and basic research.

Based on the results of these research activities, conservation and management recommendations will be made to the appropriate managment authorities for each threatened species. A particularly urgent case to be addressed is the fragmentation and degradation of forested areas occupied by the giant golden mole.

Aside from direct research and conservation activities, productive strategies might include: sending reprints of publications to people wishing to work on golden mole conservation; preparing special pamphlets on the unique aspects of golden moles and their conservation; and sending letters of concern from the IUCN to governments and conservation officials concerning specific problems associated with the conservation of these species.

Appendix 3: An Evaluation of the Status of the Elephant-Shrews *Rhynchocyon chrysopygus* and *Rhynchocyon petersi* in Coastal East Africa

Background

The mammalian Order Macroscelidea (elephant-shrews) includes 15 well-defined species that are endemic to Africa. These species are distributed in habitats that range from deserts to forests, but the majority occupy woodland and bushland habitats. The genus *Rhynchocyon* is found in lowland and montane forests in central and eastern Africa. The three species of *Rhynchocyon* are the largest and most spectacular of the elephant-shrews, with total lengths of about 52 cm, and weights of about 550g. They are diurnal and arguably among the most beautifully coloured mammals in Africa.

In the Action Plan for the African Insectivora and Elephant-shrews, it is concluded that some of the forest-dwelling species (including some forms of *Petrodromus*) may face extirpation, mainly due to habitat destruction. In assessing elephant-shrew research and conservation priorities, it has been determined that the most important research need is to develop survey techniques for the threatened forest-dwelling species, and to determine the status of the golden-rumped elephant-shrew (*Rhynchocyon chrysopygus*) and of the black-and-rufous elephant-shrew (*Rhynchocyon petersi*) in coastal East Africa. These species, especially the former, have distributions that are among the most restricted of any elephant-shrews. The golden-rumped elephant-shrew occurs only in coastal forest patches in Kenya north of Mombasa to the Somalia border, while the black-and-rufous elephant-shrew is found over a broader area, including isolated coastal forests from Mombasa south into Tanzania. Not only is the forested area along the coast relatively small, but it is also narrow, highly fragmented, and adversely affected by intense urban and agricultural development.

It is possible that *Rhynchocyon* elephant-shrews could be used as indicator species for assessing the status of primary forest habitats. There is some evidence, however, that *Rhynchocyon* elephant-shrews might occur in coastal scrub and fallow agricultural land adjacent to undisturbed forests. It is important to determine whether this is indeed the case, since it has a considerable influence on the development of conservation programmes for these unique mammals.

It is known that *Rhynchocyon* elephant-shrews are trapped and snared for food along the East African coast, but it is not known how widespread or extensive this subsistence take is, nor whether it poses a threat to the existence of these species. It is possible that some controlled hunting is an acceptable activity from a conservation perspective, especially in the context of protecting forest habitats as Biosphere Reserves, in which some controlled human activity that is compatible with the maintenance of the habitat is permitted.

Objectives

The objectives of the proposed research are:

1. to develop efficient and reliable techniques to determine the occurrence and density of *Rhynchocyon* elephant-shrews;

2. to determine whether or not the golden-rumped and black-and-rufous elephant-shrews occupy fallow agricultural land;

3. to evaluate the effect of subsistence hunting on the status of *Rhynchocyon* elephant-shrews in coastal East Africa;

4: to make arrangements and plans with authorities and colleagues in East Africa to continue and expand elephant-shrew field research;

5. to use the data obtained to assist conservation agencies in implementing appropriate field programmes.

Justification

Although the principal life history characteristics of the golden-rumped elephant-shrew were determined during a two-year field study at the Gedi Ruins, near Malindi, Kenya, there is little detailed information on the distribution, habitat preferences, and abundance of this, or other, forest-dwelling elephant-shrews. By initiating research to develop standardised methods of gathering status information for *Rhynchocyon* elephant-shrews, biologists can begin field surveys of the three species in the genus, all of which occupy isolated and potentially threatened forest habitats.

The proposed surveys in Kenya and Tanzania have been identified in this Action Plan as being among the most urgent research/conservation activities for elephant-shrews. With more reliable status information, more effective conservation strategies for *Rhynchocyon* elephant-shrews and their habitats could be developed and implemented. These activities will contribute to the conservation of some of eastern and central Africa's most unique forests, which support numerous endemic plants and animals. These forests are also important, and worth protecting as watersheds and as sources of a sustainable supply of various products, such as timber, poles, medicinal plants, and small game, not to mention tourist revenues, and recreation.

Methods

The study of the golden-rumped elephant-shrew in the forest at Gedi Ruins developed effective methods of catching and observing elephant-shrews. The animals were trapped using entangling fishing nets strung vertically in straight lines along the forest floor, and the catch per unit effect was determined. After being tagged, the animals were released unharmed. The diurnal elephant-shrews were subsequently sighted by quietly walking a grid of well-maintained trails through the forest. The density of tagged elephant-shrews was determined, as well as the sightings per unit effort, from these grid surveys. However, a more rapid method of determining elephant-shrew occurrence and abundance is needed. The observation methods used at Gedi during the 1971/72 study, along with indirect elephant-shrew signs such as sleeping nests and foraging excavations on the forest floor, will be used as the basis for developing an index of relative density using transect sampling methods. Transect techniques, however, are fraught with difficulties, especially in forested habitats. Modifications to the observation methods used in the 1971/72 Gedi Study must therefore be made. For example, transects will require visibility from the transect path to be considered. Another factor that must be addressed is the effect on elephant-shrew visibility of walking transects on unimproved trails through the leaf-littered forest floor, versus trails cleaned of litter.

The development of the transect techniques will be done at Gedi Ruins, where base-line data are available from the 1971/72 study, logistical problems are minimal, an established set of trails exists, and a relatively high density of elephant-shrews should still occur.

The standardized survey methods that are developed at Gedi will be used to assess the distribution and relative abundance of the golden-rumped and black-and-rufous elephant-shrews in coastal Kenya and Tanzania. In Kenya, these include the Diani, Shimba Hills, Arabuko-Sokoke, and Boni forests, as well as the many smaller forest patches ("kayas") between these larger reserves, In Tanzania, the forests on Mafia and Zanzibar islands and the Pugu Forest on the mainland are of particular interest. While surveying these areas, an assessment of the subsistence take of *Rhynchocyon* elephant-shrews will be made, based on the occurrence of elephant-shrew traps and snares in the forests and interviews with local residents.

Extensive areas of fallow agricultural land (mostly cashew plantations and perennial crop farms) occur in the Malindi/Gede area of coastal Kenya. These secondary habitats will be surveyed to determine the occurrence of the golden-rumped elephant-shrew and assess whether these habitats support reproducing populations.

The development of the survey techniques will be done by co-investigators familiar with elephant-shrew life history traits, and biologists from research institutions in Kenya (i.e. University of Nairobi, National Museums of Kenya, Kenya Wildlife Services) and Tanzania (University of Dar es Salaam, Mweka College of African Wildlife Management). By developing and using the proposed techniques together, the investigators will be able to carry out the proposed elephant-shrew surveys independently, and yet the results will still be comparable. Future surveys of the status of *Rhynchocyon* elephant-shrews are planned for isolated inland forests in Tanzania, Malawi, and Zambia.

Use of Results

The results obtained from the proposed research and surveys will be published in a refereed journal, such as the *African Journal of Ecology* or *Conservation Biology*. Through the participation of Tanzanians and Kenyans, the information gathered will be readily available to the government agencies responsible for developing land-use and wildlife policies.

The long-term protection of elephant-shrews and their habitats must be based on an understanding of the life history and ecology of these animals. By involving personnel from institutions in Kenya and Tanzania, it will be possible to train local biologists and students in survey techniques, and to establish interest in long-term studies of elephant-shrew behavioural ecology.

The maintenance of the biological diversity of the unique coastal forests of East Africa is of considerable local and international interest at present. For example, the National Museums of Kenya is in the process of assessing the status of the coastal forests, and the International Council for Bird Preservation is initiating a conservation programme in the Arabuko-Sokoke Forest. The information on the occurrence, distribution, and subsistence hunting of elephant-shrews can be used to assist in developing management plans for the forests. These plans would form the basis of forest conservation measures, ensuring the sustained, long-term productivity of the forests without compromising the ecological and genetic integrity of their unique flora and fauna.

Appendix 4: Members of the IUCN/SSC Insectivore, Elephant Shrew and Tree Shrew Specialist Group

Dr Martin Nicoll (Chairman)
WWF-Protected Areas, Madagascar
B.P. 738
Antananarivo 101
MADAGASCAR

Dr R. David Stone (Deputy Chairman)
10 route de Burtigny
1268 Begnins
SWITZERLAND

Dr Hisashi Abe
Hokkaido University
Institute of Applied Zoology
Faculty of Agriculture
Sapporo
JAPAN

Dr Sara Churchfield
King's College
Biosphere Sciences Division
Campden Hill Road
London W8 7AH
UNITED KINGDOM

Dr Gordon B. Corbet
27 Farnaby Road
Bromley
Kent BR1 4BL
UNITED KINGDOM

Dr Gilbert Dryden
Slippery Rock University
Slippery Rock, PA 16057
UNITED STATES OF AMERICA

Dr John F. Eisenberg
Florida Museum of Natural History
Universtiy of Florida
Gainesville, FL 32611
UNITED STATES OF AMERICA

Dr Sarah B. George
Museum of Los Angeles County
Section of Mammalogy, Natural History
900 Expostion Boulevard
Los Angeles, CA 90007
UNITED STATES OF AMERICA

Dr Edwin Gould
National Zoo
Smithsonian Institution
Washington, DC 20008
UNITED STATES OF AMERICA

Dr Graham C. Hickman
Dept of Zoology
University of Natal
Pietermaritzburg
REPUBLIC OF SOUTH AFRICA

Dr Robert S. Hoffmann
Smithsonian Institution
1000 Jefferson Drive, S.W./SI 120
Washington, DC 20560
UNITED STATES OF AMERICA

Dr Rainer Hutterer
Museum Alexander Koenig
Adenauerallee 150-164
5300 Bonn 1
FEDERAL REPUBLIC OF GERMANY

Dr Fred W. Koontz
New York Zoological Society
185th Street & Southern Blvd
Bronx, NY 10460
UNITED STATES OF AMERICA

Dr Lim Boo Liat
Malaysia Zoological Society
Dept of Wildlife & National Parks
12 Jalan Koop. Cuepacs
Taman Cuepacs, 9th mile Cheras
Kajang, Selangor 43200
MALAYSIA

Mr Nicholas Lindsay
King Khalid Wildlife Research Centre
N.C.W.C.D.
P.O. Box 61681
Riyadh 11575
SAUDI ARABIA

Mr Tiziano Maddalena
Institut de zoologie et
d'ecologie animale
Universite de Lausanne
1015 Lausanne
SWITZERLAND

Dr William J. McShea
Conservation & Research Center
National Zoological Park
Front Royal, VA 22630
UNITED STATES OF AMERICA

Prof. J. Meester
University of Natal
Department of Biology
King George V Av.
Durban 4001
REPUBLIC OF SOUTH AFRICA

Dr Patrick A. Morris
Royal Holloway and Bedford New College
Dept of Biology
Bakeham Lane, Englefield Green
Surrey TW20 9TY
UNITED KINGDOM

Dr Junaidi Payne
WWF-Malaysia
WDT No 40
89400 Likas, Sabah
MALAYSIA

Dr Jose A. Ottenwalder
Dept of Zoology, Research & Conservation
Parque Zoolique. Nac. ZOODOM/DNP
P.O. Box 2449
Santo Domingo, D.N.
DOMINICAN REPUBLIC

Ms Ana Isabela Queiroz
Servico National de Parques
Reservas e Conservacao da Natureza
Rua Filipe Folque 46 - 1o
1000 Lisboa
PORTUGAL

Dr Dioscoro S. Rabor
University of the Philippines
at Los Banos, College of Forestry
College, Laguna
PHILIPPINES

Mr Felix Rakotondraparany
Parc de Tsimbazaza
P.O. Box 4096
Antananarivo 101
MADAGASCAR

Dr Galen B. Rathbun
US Fish & Wildlife Service
P.O. Box 70
San Simeon, CA 93452
UNITED STATES OF AMERICA

Dr Bernard Richard
CESCAU
09800 Castillon
FRANCE

Dr Duane A. Schlitter
Carnegie Museum of Natural History
4400 Forbes Avenue
Pittsburgh, PA 15213
UNITED STATES OF AMERICA

Mr P.J. Stephenson
University of Aberdeen
Dept of Zoology
Tillydrone Ave
Aberdeen AB9 2TN
UNITED KINGDOM

Prof. Peter Vogel
Institut de zoologie et d'écologie animale
Universite de Lausanne
1015 Lausanne
SWITZERLAND

Dr Charles A. Woods
Florida Museum of Natural History
University of Florida
Gainesville, Fl 32611
UNITED STATES OF AMERICA

Dr Terry L. Yates
University of New Mexico
Dept of Biology
Albuquergue, NM 87131
UNITED STATES OF AMERICA

Dr C.G. van Zyll de Jong
Natural Museum of Natural Sciences
Mammalogy Section
P.O. Box 3443, Station D
Ottawa, Ont. K1P 6P4
CANADA

References

Ansell, W.F.H. 1960. Mammals of Northern Rhodesia. Government Printer, Lusaka.

Ansell, A.D.H. and Ansell, P.D.H. 1969. *Petrodromus tetradactylus* at Ngoma. *Puku* 5: 211-213.

Ansell, W.F.H. and Ansell, P.D.H. 1973. Mammals of the northeastern montane areas of Zambia. *Puku* 7: 21-69.

Ansell, W.F.H. and Dowsett, R.J. 1988. Mammals of Malawi. The Trendrine Press, St. Ives, Cornwall, England.

Aulagnier, S. and Thevenot, M. 1986. Catalogue des mammiferes Sauvages du Maroc. *Travaux de l'Institut Scientifique, Rabat. Serie Zoologie* No. 41.

Brown, J.C. 1964. Observations on the elephant shrews (Macroscelididae) of Equatorial Africa. *Proc. Zool. Soc. Lond.* 143: 103-119.

Brown, L.H. 1981. The conservation of forest islands in areas of high human density. *Afr. J. Ecol.* 19: 27-32.

Coetzee, C.G. 1969. The distribution of mammals in the Namib Desert and adjoining inland escarpment. *Scient. Pap. Namib Des. Res. Stn.* 40: 23-36.

Collins, N.M. and Clifton, M.P. 1984. Threatened wildlife in the Taita Hills. *Swara* 7: 10-14.

Collar, N.J. and Stuart, S.N. 1985. Threatened birds of Africa and related islands. The ICBP/IUCN Red Data Book, Part I. ICBP, Cambridge, UK, and IUCN, Gland, Switzerland.

Corbet, G.B. 1971. Family Macroscelididae. Part 1.5, pp 1-6 in J. Meester and H.W. Setzer (eds.). The Mammals of Africa: An Identification Manual. Smithsonian Inst. Press, Washington, D.C.

Corbet, G.B. and Hanks, J. 1968. A revision of the elephant-shrews, Family Macroscelididae. *Bull. Brit. Mus. Nat. Hist. Zool.* 16: 1-111.

Corbet, G.B. and Neal, B.R. 1965. The taxonomy of the elephant shrews of the Genus *Petrodromus*, with particular reference to the East African coast. *Rev. Zool. Bot. afr.* 71: 49-78.

Crandall, L.S. 1964. The Management of Wild Mammals in Captivity. Univ. Chicago Press, Chicago.

Critch, P.J. 1969. The rock elephant shrew. *African Wildlife* 23: 139-144.

Dachsel, M. 1964. Die Atlantische Elefantenspitzmaus. *Aquarien und Terrarien Zeitschrift* 17: 217-219.

Diamond, A.W. 1981. Reserves as oceanic islands: lessons for conserving some East African montane forests. *Afr. J. Ecol.* 19: 21-25.

Eisenberg, J.F. and Gould, E. 1970. The tenrecs: a study in mammalian behavior and evolution. *Smithson. Contrib. Zool.* 27: 1-127.

Flower, S.S. 1931. Contributions to our knowledge of the duration of life in vertebrate animals. 5. *Mammals. Proc. Zool. Soc. Lond.* Pp. 145-234.

Grandidier, G. and Petit, G. 1931. Un type nouveau de Centitédé malgache, *Paramicrogale occidentalis*. *Bull. Soc. Zool. France* 56: 126-139.

Goodwin, H.A. and Goodwin, J.M. 1973. List of mammals which have become extinct or are possibly extinct since 1600. IUCN Occ. Pap. 8.

Gould, E. and Eisenberg, J.F. 1966. Notes on the biology of the Tenrecidae. *J. Mamm.* 47: 660-686.

Heim de Balsac, H. 1954. Un genre inédit et inattendu de mammifère (Insectivore Tenrecidae d'Afrique Occidentale). *C.R. Acad. Sci. Paris.* 239: 102-104.

Heim de Balsac, H. 1968. Considerations preliminaires sur le peuplement des montagnes africaines par les Soricidae. *Biol. Gabon.* 4: 299-323.

Heim de Balsac, H. and Hutterer, R. 1982. Les Soricidae (Mammiferes Insectivores) des iles du Golfe de Guinee: faits nouveaux et problemes biogeographiques. *Bonn. zool. Beitr.* 33: 133-150.

Hedberg, I. and Hedberg, O. (eds.). 1968. Conservation of vegetation in Africa south of the Sahara. *Acta Phytogeographica Suedica* 54.

Hoesch, W. 1959. Zur Jugendentwicklung der Macroscelididae. *Bonn. zool. Beitr.* 3/4: 263-265.

Hoogstraal, H. 1950. Mission: Malaria. *Nat. Hist.* 59: 104-111, 140-141.

Hoopes, P.J. and Montali, R.J. 1980. Tail lesions in captive elephant shrews. Pp. 425-430 in R.J. Montali and G. Migaki (eds.). The Comparative Pathology of Zoo Animals. Smithsonian Instututution, Washington, D.C.

Horst, C.J. van der. 1946. Some remarks on the biology of reproduction in female of *Elephantulus*, the holy animal of set. *Trans. Roy Soc. S. Afr.* 31: 181-199.

Horst, C.J. van der. 1954. *Elephantulus* going into anoestrus: menstruation and abortion. *Phil. Trans. Roy. Soc. Lond.* 238B: 27-61.

Houck, J., C. Lewis and Smith, R. 1981. Captive husbandry and propagation of East African long-eared elephant shrews (*Elephantulus rufescens*). Pp. 79-84 in J. Mellen and A. Littlewood (eds.). Recent Developments in Research and Husbandry at the Washington Park Zoo. Washington Park Zoo, Portland, Oregon.

Howell, K.M. 1981. Pugu Forest Reserve: biological values and development. *Afr. J. Ecol.* 19: 73-81.

Hutterer, R. 1981. *Crocidura manengubae* n. sp. (Mammalia: Soricidae), eine neue Spitzmaus aus Kamerun. *Bonn. zool. Beitr.* 32: 241-148.

Hutterer, R. 1986. Synopsis der Gattung *Paracrocidura* (Mammalia: Soricidae), mit Beschreibung einer neuen Art. *Bonn. zool. Beitr.* 37: 73-90.

Hutterer, R. and Dippenaar, N.J. 1987. A new species of *Crocidura* Wagler, 1832 (Soricidae) from Zambia. *Bonn. zool. Beitr.* 38: 1-7.

Hutterer, R. and Happold, D.C.D. 1983. The shrews of Nigeria (Mammalia: Soricidae). *Bonn. zool. Monogr.* 18: 1-79.

Hutterer, R., Van der Straeten, E. and Verheyen, W.N. 1987. A checklist of the shrews of Rwanda and biogeographical considerations on African Soricidae. *Bonn zool. Beitr.* 38: 155-172.

Hutterer, R. and Verheyen, W.N. 1985. A new species of shrew, genus *Sylvisorex*, from Rwanda and Zaire (Insectivora, Soricidae). *Z. Saugetierkunde* 50: 266-271.

Jenkins, P.D. 1988. A new species of *Microgale* (Insectivora: Tenrecidae from northeastern Madagascar. *Am. Mus. Novitates* 2910: 1-7.

Keppel, A. 1985. *Elephantulus rufescens* at the National Zoological Park. Animal Keepers' Forum, Special Edition 1985: 392-396.

Kingdon, J. 1971. East African Mammals: An Atlas of Evolution in Africa. Vol. I. Academic Press, London.

Kingdon, J. 1974. East African Mammals: An Atlas of Evolution in Africa. Vol IIA. Academic Press, London.

Kingdon, J. 1981. Where have the colonists come from? A zoogeographical examination of some mammalian isolates in eastern Africa. *Afr. J. Ecol.* 19: 115-124.

Koontz, F.W. 1984. Sternal gland scent communication in the rufous elephant-shrew, *Elephantulus rufescens* Peters, with additional observations on behavior and reproduction in captivity. Ph.D. Dissertation, University of Maryland.

Kuhn, H. 1971. An adult female *Micropotamogale lamottei*. *J. Mammal.*

52: 477-478.

Lacy, R.C. 1988. A report on population genetics in conservation. *Conservation Biology* 2: 245-247.

Lovett, J. 1985. Moist forests of eastern Tanzania. *Swara* 8: 8-9.

Lovett, J.C. and Thomas, D.W. 1988. Report on a visit to Kanga Mountain, Tanzania. *E. Afr. Nat. Hist. Bull.* 18: 19-22.

Lumpkin, S. 1986. The elephant-shrew...by a nose! Zoogoer (National Zoological Park, Wash., D.C.) 15: 8-10.

Lumpkin, S. and Koontz, F.W. 1986. Social and sexual behavior of the rufous elephant-shrew (*Elephantulus rufescens*) in captivity. *J. Mamm.* 67: 112-119.

Lumpkin, S., Koontz F., and Howard, J.G. 1982. The oestrous cycle of the rufous elephant-shrew, *Elephantulus rufescens*. *J. Reprod. Fert.* 66: 671-673.

MacPhee, R.D.E. 1987. The shrew tenrecs of Madagascar: systematic revision and Holocene distribution of *Microgale* (Tenrecidae, Insectivora). *Am. Mus. Novitates* 2889: 1-45.

McKenna, M.C. 1975. Towards a phylogenetic classification of the Mammalia. In: Luckett, W.P. and Szalay, F.S. (eds.), Phylogeny of the Primates. A multidisciplinary approach. Plenum Press, New York.

Malzy, P. 1965. Un mammifère aquatique de Madagascar le Limnogale. *Mammalia* 29: 399-411.

Meester, J. 1976. South African Red Data Book: Small mammals. *South African National Scientific Programmes Report* 11, Pretoria, Council for Scientific and Industrial Research.

Meester, J. and Dippenaar, N.J. 1978. A new species of *Myosorex* from Knysna, South Africa (Mammalia: Soricidae). *Ann Transvaal Mus.* 31: 29-42.

Meester, J. and Setzer, H.W. 1974. The mammals of Africa: an identification manual. Smithsonian Institution Press, Washington, D.C.

Meester, J.A.J., Rautenback, I.L., Dippenaar, N.J. and Baker, C.M. 1986. Classification of Southern African mammals. *Transvaal Museum Mongraph* No. 5.

Moomaw, J.C. 1969. A study of the plant ecology of the coast region of Kenya Colony. Government Printer, Nairobi, Kenya.

Moreau, R.E. 1966. The bird faunas of Africa and its islands. Academic Press, London.

Nicoll, M.E. 1985. The biology of the Giant otter-shrew *Potamogale velox*. *Nat. Geog. Soc. Res. Rep.* 21: 331-337.

Nicoll, M.E., Rakotondraparany, F. and Randrianasolo, V. 1988. Diversité des petits mammifères en forêt tropicale humide de Madagascar, analyse préliminaire. In: Rakotovao, L., Barre, V. and Sayer, J. (eds.), L'Equilibre des ecosystemès forestiers à Madagascar: Actes d'un séminaire international, pages 241-252. IUCN, Gland, Switzerland and Cambridge, United Kingdom.

Nowak, R.M. and Paradiso, J.L. 1983. Walker's mammals of the world. Johns Hopkins University Press, Baltimore and London.

Olney, P.J.S. 1979. Species of wild animals bred in captivity during 1977 and multiple generation captive births. *Internat'l. Zoo Yrbk.* 19: 297-383.

Patterson, B. 1965. The fossil elephant shrews (Family Macroscelididae). *Bull. Mus. Comp. Zool.*, Harvard Univ. 133: 295-335.

Pocock, R.I. 1912. On elephant-shrews. *Proc. Zool. Soc. Lond.* 1912: 142-144.

Prigogine, A. 1985. Conservation of the Avifauna of the forests of the Albertine Rift. *ICBP Technical Publ.* 4: 277-294.

Rahm, U. 1966. Les mammifères de la forêt équatoriale de l'est du Congo. *Ann. Mus. Roy. Afr. Cet. (Tervuren)*, ser. 8. 149: 39-121.

Rankin, J.J. 1965. Notes on the ecology, capture and behaviour in captivity of *Nasilio brachyrhynchus*. *Zool. Afr.* 1: 73-80.

Rankin, J.J. 1967. The transport and laboratory maintenance of the elephant-shrew, *Nasilio brachyrhynchus* (A. Smith). *J. Inst. Animal Technicians* 18: 178-183.

Rathbun, G.B. 1979. The social structure and ecology of elephant-shrews. *Z. Tierpsychol.* suppl. 20: 1-77.

Rathbun, G.B., Beaman, P. and Maliniak, E. 1981. Capture, husbandry and breeding of rufous elephant-shrews, *Elephantulus rufescens*. *Internat'l. Zoo Yrbk.* 21: 176-184.

Rautenbach, I.L. 1978. The mammals of the Transvaal. Ph.D. Thesis, University of Natal.

Rodgers, W.A. and Homewood, K.M. 1982a. Biological values and conservation prospects for the forests and primate populations of the Uzungwe Mountains, Tanzania. *Biol. Conserv.* 24: 285-304.

Rodgers, W.A. and Homewood, K.M. 1982b. Species richness and endemism in the Usambara mountain forests, Tanzania. *Biol. J. Linn. Soc.* 18: 197-242.

Rodgers, W.A., Owen, C.F. and Homewood, K.M. 1982. Biogeography of East African forest mammals. *J. Biogeog.* 9: 41-54.

Rosenthal, M. 1975. The management, behavior and reproduction of the short-eared elephant shrew. M.A. Thesis, Northeastern Illinois University.

Sauer, E.G.F. and Sauer, E.M. 1972. Zur Biologie de kurzohrigen Elefantenspitzmaus. *Zeitschrift des Kolner Zoo* 4: 119-139.

Sequignes, M. 1983. La torpeur chez *Elephantulus rozeti* (Insectivora, Macroscelididae). *Mammalia* 47: 87-91.

Silkiluwasha, F. 1981. The distribution and conservation status of the Zanzibar red colobus. *Afr. J. Ecol.* 19: 187-194.

Smithers, R.H.N. 1971. The Mammals of Botswana. *National Museums of Rhodesia, Museum Memoir* No. 4.

Smithers, R.H.N. 1983. The Mammals of the Southern African Subregion. Univ. Pretoria, Pretoria, Republic of South Africa.

Smithers, R.H.N. 1986. South African Red Data Book - Terrestrial Mammals. South African National Scientific Programmes Report No. 125.

Struhsaker, T. 1981. Forest and primate conservation in East Africa. *Afr. J. Ecol.* 19: 99-114.

Stuart, S.N. 1981. A comparison of the avifaunas of seven east African forest islands. *Afr. J. Ecol.* 19: 133-151.

Stuart, S.N. 1985. Rare forest birds and their conservation in eastern Africa. *ICBP Technical Publ.* 4: 187-196.

Thornback, J. and Jenkins, M. 1982. The IUCN Mammal Red Data Book. IUCN, Gland, Switzerland.

Tripp, H.R.H. 1972. Capture, laboratory care and breeding of elephant-shrews (Macroscelididae). *Laboratory Animals* 6: 213-224.

Van Valen, L. 1967. New Paleocene insectivores and insectivore classification. *Bull. Amer. Mus. Nat. Hist.* 135: 217-284.

Vogel, P. 1983. Contribution à l'écologie et la zoogéographie de *Micropotamogale lamottei* (Mammalia: Tenrecidae). *Rev. Ecol. (Terre et Vie)* 38: 37-49.

Walker, E.D. 1955. African elephant-shrews. *Nature* 48: 295-297, 332.

White, F. 1983. *The vegetation of Africa*. UNESCO, Paris.